# Bred for
# Perfection

PUBLISHING FOR THE WORLD
125 Years

THE JOHNS HOPKINS UNIVERSITY PRESS

ANIMALS, HISTORY, CULTURE

*Harriet Ritvo, Series Editor*

# Bred for
# Perfection

SHORTHORN CATTLE, COLLIES,
AND ARABIAN HORSES
SINCE 1800

∞

*Margaret E. Derry*

The Johns Hopkins University Press
BALTIMORE AND LONDON

© 2003 The Johns Hopkins University Press
All rights reserved. Published 2003
Printed in the United States of America on acid-free paper
2 4 6 8 9 7 5 3 1

The Johns Hopkins University Press
2715 North Charles Street
Baltimore, Maryland 21218-4363
www.press.jhu.edu

Library of Congress Cataloging-in-Publication Data
Derry, Margaret Elsinor, 1945–
Bred for perfection : shorthorn cattle, collies, and arabian horses
since 1800 / Margaret E. Derry.
p. cm. — (Animals, history, culture)
Includes bibliographical references (p.  ) and index.
ISBN 0-8018-7344-4 (hardcover : alk. paper)
1. Shorthorn cattle—Breeding—History—19th century.
2. Collie—Breeding—History—19th century.   3. Arabian
horse—Breeding—History—19th century.   I. Title.   II. Series.
SF199.S56 D47 2003
636.08′2′09—dc21
2002015859

A catalog record for this book is available from the British Library.

*For my husband, Douglas L. Derry, F.C.A.*

# CONTENTS

# PREFACE

"The best [purebred] animals in the world have always been bred for the love of them or the love of breeding and caring for them, rather than purely for profit," wrote a mid-twentieth-century Collie expert.* Do purebred breeders really produce animals more for the joy of breeding than for the money they generate? This book attempts to answer that question, and others as well, by examining the history, over two centuries in Britain and North America, of three breeds: Shorthorn cattle, Collie dogs, and Arabian horses. The subject of animal breeding is approached from its center; that is, breeding practices in relation to their supportive structures. The book outlines what breeders did in actual breeding programs and puts those accounts into a historical context to help readers understand patterns in purebred breeding, how markets relate to breeding programs, and how a British-American connection could be so significant to the whole system. I hope to demystify the underlying mechanics of modern purebred breeding, giving lay readers an overview of the subject. Purebreds illustrate how selection by humans can mold animals. But while most people understand the idea of breed purity and its relation to the manipulation of life, many have little appreciation of how purebred breeding works or what drives it.

Modern purebred breeding lies in some zone between science and culture; both influenced the development of the method. While the book indicates the complexity of the system's relation to science *and* culture, I concentrate more on how purebred breeding was and is practiced and on the way its supportive structures developed. I hope, though, that seeing how purebred breeding functions will make some readers want to pursue the former topic in more depth. We could, for example, understand better what aspects of the method evolved solely from certain cultural environments and why, even today, many of these culturally based features wear

* M. Denlinger, *The Complete Collie*, 3d ed. (Richmond, Va.: Denlinger's, 1949), 9.

a scientific mask. The subject is rich, and beyond the history of animals' relations with humanity, it can do much to explain the way human knowledge and perceptions about the world generally have evolved over time.

The breeding method I discuss began its development in late eighteenth-century Britain. Earlier efforts at "pure" breeding (producing, for example, Pekinese dogs in China, Arabian horses in Arabia, and Greyhounds in Egypt)* did not share certain critical features with this British endeavor. First, even in the beginning, the new method was far more widely used. It affected many breeding operations, not a few isolated ones, and many breeds within different species. Second, it established a breeding procedure that in the end produced much greater variation in many species than had been seen before. Third, it linked pedigree keeping to breeding methodology in a novel but important way: by a public registry. At various times before the eighteenth century, some individual breeders kept track of the ancestry of animals they bred themselves, but these were isolated personal records. (Early Arabian horse breeding was an exception to this pattern; animal genealogy was preserved over generations by word of mouth.) Public pedigrees collected in writing the ancestry of animals belonging to a "breed" from many breeding programs over extended periods. Extensive pedigree keeping of this nature allowed for the development of more clearly defined breed types and led to a global spread of breed varieties. The system resulted in the creation of highly marketable animals, certified to be true to a clearly specified breed type based on registered pedigrees that record the animals' ancestral background in public studbooks.

In eighteenth-century Britain, people began to experiment with selective breeding ideas, used in various forms at different times in many parts of the world. Through this experimentation, there developed a formalized and publicized procedure with established rules that could be used to create new "breeds." When the method became connected to public record keeping in Britain in the early nineteenth century, a new purebred breeding system was born. A particularly significant trade in "purebred" animals evolved between Britain and the United States during the nineteenth century, and that international market would mold the system's entire structure. Because certain patterns specific to Shorthorn breeding and transatlantic markets from the 1830s until 1900 shaped the way all purebred breeding developed, an analysis of Shorthorn affairs is critical.

* J. Clutton-Brock, *A Natural History of Domesticated Animals*, 2d ed. (Cambridge: Cambridge University Press, 1999), 47.

Within the Shorthorn world, for example, there appeared the earliest attempts at regulating international markets through pedigrees.

I provide narratives about historical Shorthorn production that explain
how certain patterns evolved: why British nobles came to value the genetics of Hubback, a bull from an unknown background that a farmer found
grazing by the roadside; what made early nineteenth-century cattlemen in
Ohio country purchase particular stock in Britain; and why an Englishman
would spend over $40,000 to buy a seven-year-old red-and-white cow,
named 8th Duchess of Geneva, in the United States. This transatlantic
trade affected breeding strategies in other countries and revealed, for example, why Canadian farmers shifted their agricultural operations in the
1890s simply because a certain bull, Roger, was related to Shorthorns imported into the United States in 1817.

Events in the Shorthorn world shaped and consolidated a structure for
the new purebred system, regulating the breeding of animals even more
effectively. At this stage modern purebred breeding became connected
with the reproduction of animals based purely on beauty or fancy points.
By looking at the purebred dog fancy from the late nineteenth century to
the late twentieth, we see how the method supported the "sport" of breeding. Because Collies were one of the first breeds to receive attention when
selection for canine beauty evolved in Britain (and then in North America), early characteristics of the international dog fancy are apparent in
Collie history. Central to the dynamics of the Collie fancy, as with Shorthorns, was the market between Britain and the United States. But market
questions could be related to other issues in complex ways. For example,
when the wealthy financier J. P. Morgan decided in 1906 to compete with
another New Yorker, Samuel Untermyer, for the right to buy a high-priced
British-bred Collie named Squire of Tytton, Morgan might not have been
interested solely in a good investment. He valued another expensive Collie primarily as a pet: Sefton Hero, a beautiful gold-and-white champion
from Britain, was Morgan's companion and regularly slept under his bed.
Collie breeding shows that single animals could dictate the future of a
breed. Wishaw Clinker, a Collie born in the 1890s, came to represent
deeply embedded conflicts in Collie breeding that are still with us today.
Another British dog, Anfield Model, influenced the direction of Collie
fortunes in the United States over the twentieth century.

Shorthorns and Collies originated in Britain through modern breeding
methods. To determine how the significant breeding and marketing patterns found in those two breeds could become pervasive in world animal
breeding, I assess what happened when American demand arose for an an-

imal not created under the system or in Britain. Looking at Arabian horse breeding reveals how important these ubiquitous underlying characteristics could be to the dynamics of global purebred breeding. The Arabian existed in fixed type as a breed when it left its Eastern homeland, long before any Westerners began to produce "purebred" Arabians. The horses would become part of a world trade through markets that originally existed between Britain and the United States. American breeders imported Arabians bred in Britain, but they also sought the genetics of British-bred animals in other countries. Ultimately, however, American breeders would look for Arabians from the national breeding programs of many countries. The Arabian horse story leads not just to the breeding of Arabians in Britain and the United States, but also to their history in Arabia, Egypt, Poland, Russia, and Spain. American demand for the horses bred in these countries ultimately created a modern global Arabian world. Breeders in Germany, Holland, Sweden, Australia, Canada, Brazil, and Argentina followed Americans in buying Egyptian, Polish, Russian, and Spanish horses. A boom in Arabian horses developed, fueled by American tax laws that encouraged buying the horses at very high prices.

Stories about Arabian horse breeding in this book help us understand how global production of Arabians led to curious patterns, which sometimes caused difficulties. The vision of one English couple, Wilfrid and Lady Anne Blunt and their daughter, Lady Wentworth, was able to change and at the same time preserve ancient breeding of Arabians from the East, then ultimately to spread their genetics to the world. In the process one stallion, Skowronek, born in Poland in 1908, came to influence the breeding of Arabians everywhere in the world and to affect international pedigrees. Pedigree culture could dictate horse breeding, and it explained why purebred Arabians born in Spain were once considered unacceptable in the world market and why a particular stallion, Kurdo III, was blamed for the presence of impure Arabians in South America. Breeders faced serious questions. What made an Arabian pure by Eastern standards? What made an Arabian authentic to original type? How did improvement relate to preservation of original type? Could pedigrees preserve purity?

A few other comments. First, the stories I tell here involve specific animals as well as breeders and their programs for specific reasons. One must appreciate the ideals of the breeders, which were particularly evident in the animals they produced. Certain aspects of breeder ideology can be understood in no other way than through the animals. Second, to make spelling consistent, I have conformed throughout to American usage, even within quotations. And third, the book relies heavily on breeder testimony, which emanated from specialist journals and books relating to specific breeds. The material offered two types of information. One related to facts such as dates, outlines of breed standards and rules for pedigrees, and the part governments played in purebred affairs. The material also revealed breeders' opinions and contemporary biases and therefore provided critical insight into what they thought over time. For example, a book written in the 1940s on the care and history of Collies stated, "Castrated dogs are monstrosities, even worse than spayed bitches. The horrible practice of castration is mentioned here only to say it should never be done."* Today most veterinarians recommend castrating all males that will not be used for breeding.

Literature on various aspects of purebred breeding has enriched this book. Humanist scholars (particularly geographers and historians) have studied this breeding method to explore both the spread of animal improvement in the modern age and ideas about the relation of animal society to human society. John Walton, a geographer, studied the evolution of purebred livestock and its effect on the overall improvement of farm animals. He concentrated on the breeding of cattle and sheep and assessed how much improved general cattle and sheep production resulted from using the new system.** Harriet Ritvo, a historian, looked at how the method and attitudes toward animals generally dovetailed with views on human culture. She made it clear that anthropomorphism colored human vision about what constituted a modern, improved creature.† Historians

---

* Denlinger, *Complete Collie*, 72.

** J. Walton, "The Diffusion of Improved Shorthorn Cattle in Britain during the Eighteenth and Nineteenth Centuries," *Transactions of the Institute of British Geographers*, n.s., 9 (1984): 22–36; "Pedigree and Productivity in the British and North American Cattle Kingdoms before 1930," *Journal of Historical Geography* 25 (1999): 441–62; "Pedigree and the National Herd circa 1750–1950," *Agricultural History Review* (henceforth *AHR*) 34 (1986): 149–70; and "The Diffusion of Improved Sheep Breeds in Eighteenth- and Nineteenth-Century Oxfordshire," *Journal of Historical Geography* 9 (1983): 175–95. For information on purebred cattle breeding and improvement of general herds see M. Derry, "The Development of a Modern Agricultural Enterprise: Beef Cattle Farming in Ontario, 1870–1924" (Ph.D. diss., University of Toronto, 1997); M. Derry, *Ontario's Cattle Kingdom: Purebred Breeders and Their World, 1870–1920* (Toronto: University of Toronto Press, 2001).

† H. Ritvo, *The Animal Estate* (Cambridge: Harvard University Press, 1987); H. Ritvo, *The Platypus and the Mermaid and Other Figments of the Classifying Imagination* (Cambridge: Harvard University Press, 1997); H. Ritvo, "Race, Breed and Myths of Origin: Chillingham Cattle as Ancient Britons," *Representations* 39 (1992): 3–22.

of agriculture have also addressed the topic, explaining the system within the world of farming.* I enlarge on the excellent material found in all these sources by discussing breeding theory in more detail and in relation to actual animals, by linking the work of breed associations to matters of government, and by explaining how pedigree standards drove breeding programs in relation to markets.

The breeding system under discussion is, of course, only one aspect of the larger issue of human relationships with all animals, and scholars from various other fields—agricultural scientists, animal biologists, and scholars concerned with domestication being a few—have also touched on the subject. Usually they do so in connection with their specialty or with the larger topic of general animal improvement. They normally do not analyze the dynamics of the breeding method described here, and they tend to refer to it only in such general terms as selective breeding, regulated by breeder associations.** They make little effort to separate the meaning and implications of private and public pedigrees or to show how significant it could be to link public pedigree recording to certain identifiable breeding practices and markets. But scholars specializing in these areas have done much interesting work on general animal improvement and the implications of major movements like domestication.

I find the whole topic of purebred breeding interesting for practical, artistic, and academic reasons. Fifteen years of breeding, showing, and selling purebred beef cattle has taught me something of the internal world of purebred animal breeding and the dynamics behind purebred breeding practices. The animals interest me on another level as well. They are subjects for my paintings, as I try to capture on canvas the special meaning that animals have for humanity. I want to portray the ancient, unconscious bond between domestic creatures and people. The art finds buyers, so I must be conveying something of my feelings. My involvement with purebred breeding on practical and artistic levels led me to study the evolution

* See R. Trow-Smith, *A History of British Livestock Husbandry, 1700–1900* (London: Routledge and Kegan Paul, 1959); M. Lerner and H. Donald, *Modern Developments in Animal Breeding* (New York: Academic Press, 1966), 155–86; and A. Fraser, *Animal Husbandry Heresies* (London: Crosby Lockwood, 1960), 21–31, 33–49, 51–67, 69–89.

** See, for example, R. Coppinger and L. Coppinger, *Dogs* (New York: Scribner, 2001), 126, 244–45, 250–51; Clutton-Brock, *Natural History*, 40–48; T. Grandin, ed., *Genetics and the Behaviour of Domestic Animals* (London: Academic Press, 1998); J. Lush, "Notes on Animal Breeding," manuscript, 1933; and J. Lush, *Animal Breeding Plans* (Ames, Iowa: Collegiate Press, 1937).

of the present system. My work as an academic historian led quite naturally to the topic, for most of my earlier research had been devoted to agriculture, and more particularly to the role that improved (in this case purebred) livestock played in modern agricultural developments in the Western world.

As I attempted to explain the internal world of purebred breeders and breeding based on my lifelong engagement with animals, history, and art, throughout my writing I had wonderful help. My connections with the University of Guelph and the University of Toronto gave me a resource base, and many people came to my aid over specific issues. At the Pierpont Morgan Library in New York, Christine Nelson helped me find information on J. P. Morgan's dogs. After my visits to New York, she notified me whenever she found more Collie pictures. At the American Kennel Club library in New York, Barbara Kolk gave me a great deal of information and made my visits very productive. Librarians at the Kennel Club in London, England, and the Library of Congress in Washington, D.C., eased my research. Bill Bawden of PricewaterhouseCoopers helped me find information on the American tax law that so influenced Arabian horse breeders. My daughter Alison, a Ph.D. candidate in biology, supplied me with some interesting scientific information. When I had completed a draft, my son David, a recent masters graduate in tropical agriculture, and my husband, Douglas Derry, took a great deal of time to read the manuscript. Both offered valuable suggestions. Peter Montgomery, a good friend, also commented on it.

Harriet Ritvo told me about the series Animals, History, Culture and encouraged me to send the manuscript to the Johns Hopkins University Press. I am very thankful that she did, and her advice has been invaluable. She provided many insightful comments that vastly improved the work. I have always admired her impressive knowledge of the history and culture of animals in relation to people. Suggestions from unnamed readers also proved very useful. Robert J. Brugger, acquisitions editor at the Press, gave me ideas that helped me improve the book's structure. Melody Herr and Julie McCarthy aided me through the process of publication. Alice Bennett did much to make this a better-written book. I extend my sincere thanks to all these people.

The purebred breeding world was (and is) a complicated one, made up of many interconnected stories. While this book deals with only a few of these accounts, I hope to demonstrate that with sufficient analysis, even a small number of narratives can explain a complex system. Because of the emphasis on actual breeders and animals, a sense of theater emerges in the stories that follow. They confirm that passion, greed, idealism, and love played a role in the world of breeders, and that certain international trade regulations were central to the evolution of an animal industry. I hope that this review of modern purebred breeding will enrich readers' understanding of how animals have fit into human society over the past two hundred years and will clarify what motivated and still motivates those who breed purebred animals.

# Bred for
# Perfection

# One

# DEVELOPING A MODERN METHOD
# OF PUREBRED BREEDING

This chapter introduces the complex world of modern purebred breeding by explaining how fundamental aspects of the system evolved from the late eighteenth century to the late twentieth and also how larger movements such as genetics and eugenics influenced the production of purebred animals. The historical overview given in the chapter demonstrates how the system was affected by interacting forces and, by showing the way they fit into a larger perspective, provides context for the stories about Shorthorns, Collies, and Arabians that follow in subsequent chapters.

Animal breeding is an ancient human occupation, dating back to the time of domestication at least fourteen thousand years ago. Some scholars argue that even the earliest patterns of human association with other living things arose out of the need to achieve better agricultural production.[1] Whether or not agricultural demands triggered human interest in manipulating other species, domestication did from the beginning introduce new breeding patterns for life under human control. The animals became cut off from the larger gene pool of their species. This isolation occurred in two phases, both of which encouraged inbreeding. At first, people separated the animals they wanted to use from the wild population but allowed the creatures so selected to breed randomly. Later, humans controlled the breeding of their charges.[2] Over time the animals came to have less acute sight and hearing and smaller brains and to retain juvenile characteristics into adulthood (an effect called neoteny). In dogs and horses the teeth became more crowded and the heads shorter. Most domesticated creatures attained a smaller adult size than their wild counterparts, which promoted earlier maturity and greater fertility.[3]

The continued artificial selection of animals for breeding led not only to marked physical changes from their wild progenitors but also to distinctive types. Varieties of dogs, cattle, and horses evidently existed by four thousand years ago.[4] By the nineteenth century more varieties within these species had come into being in many parts of the world. But scholars em-

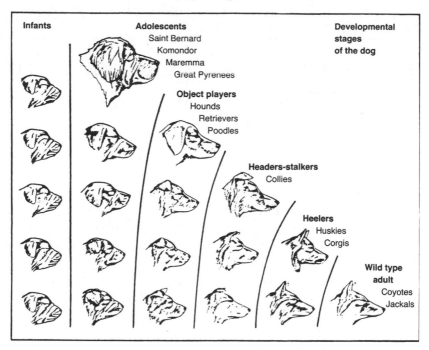

Neoteny in dogs. Domestication caused tamed animals to retain juvenile features into adulthood. The rows show how much the heads of various canines diverge from a common infant shape as they grow to adulthood. Note that a greater change takes place in coyotes and jackals than in any breed of dog. Illustration by C. Lyon, *Smithsonian Magazine.*

phasize that in spite of these developments, selection remained very haphazard and slow. As Juliet Clutton-Brock recently wrote, "The production and retention of favored characteristics through several generations was very much a hit-and-miss affair and remained so from the time of the ancient Egyptians until the publication of *The Origin of Species* by Darwin in 1859."[5]

New attitudes toward all living things, however, had been slowly crystallizing in Britain, and they laid the groundwork for more rapid changes in animal variation. In the early modern period, three particular features can be identified in that overarching shift. As Keith Thomas stated: "The early modern period [in Britain saw] the elimination of many wild animals, the increased exploitation of domestic ones, and a rise in interest in [a] third category, the pet."[6] Plant life would be categorized in the same way. People destroyed noxious wild plants, raised domesticated crops for food, and nurtured "tame" flowers for pleasure.[7] The move to mold domesticated life (as opposed to wildlife, normally viewed as something to be contained, observed, or eliminated) gathered momentum in the late eighteenth century.[8] The manipulation of plants and the manipulation of animals also became increasingly intertwined. Creating better fodder crops for animals, for example, commanded attention at least as much as improving plants for human consumption. One tool used to achieve change in both plants and animals was hybridizing—crossing species that could interbreed. While hybridizing might be seen as a tool altering characteristics for the better, it could also be viewed as bringing about degeneration.[9] The meaning of hybridizing in relation to both improvement for desired characteristics and also degeneration would later become blurred in many people's minds.[10] Some would confuse crossing of breeds with hybridizing between species. The greater variation within species that resulted from more selective breeding, but without hybridizing, would make it hard for many animal breeders to define "breed" or to understand how either breed or hybridizing related to the word "species."[11] To some degree it has been hard ever since for breeders to define "breed," within that framework.[12]

In the late eighteenth century, a number of British breeders attempted to "improve" horses, sheep, and cattle, and in the process they enhanced deviation within species. The most famous of these breeders was Robert Bakewell, and he commonly received the credit for all their work (both over his own lifetime and also throughout the nineteenth century). He was also credited with making intense inbreeding over generations acceptable and with convincing people that the method could create distinct breeds. Specifically, Bakewell tried to enhance the strength of the Shire horse, the meat of the Leicester sheep, and the beefiness of the Longhorn cow. He used a formal method of intense inbreeding, mating closely related animals over many generations. He also used a weaker version of inbreeding (line breeding) in which the animals were less closely related, but he avoided outbreeding (breeding unrelated individuals). In essence, by formalizing a set system for inbreeding, Bakewell established a recognizable procedure for making breeds. While his practical work in the end left few long-term effects on the breeds he attempted to change (except the Leicester sheep), the method ascribed to him survived and was used successfully by other breeders to create new, lasting modern breeds.

Bakewell's work became popularized through the writings of the agriculturist Arthur Young, which helped spread information—and favorable commentary—on the system. Bakewell also knew how to market his breeding, and he sold his breeding ideas through his livestock. By creating a demand for his animals, he linked monetary value to the very idea of breed and its contingent notion of improvement. One scholar assessing Bakewell's work claimed that the famous breeder created the demand for his animals, then limited the supply in order to increase their price. Thus, this scholar said, the new idea of "breed" was from the beginning nothing more than "an ingenious marketing and publicity mechanism."[13] The market, it seems, played a critical early role in formalized selective breeding.

Three significant principles of Bakewell's work passed down to emerging breeders and influenced the direction purebred breeding took over the next century. First, that new breeds could be created and then maintained by an intensive and well-defined system of inbreeding. Second, that the stock belonging to "breeds" that resulted from this breeding method had economic value. And third, that type fixing of breeds should be done through male lines. This third tenet of Bakewell's system, which gendered breeding methods, evolved out of another and more important aspect of his work. He believed that real excellence could be demonstrated by what today we call progeny testing. Animals (and males were the main focus) should be judged not by either their individual value or their hereditary background, but by the excellence of their progeny. Bakewell lent out bulls, stallions, and rams to other farmers and then assessed what these sires produced. Based on the descendants of each, Bakewell came to an opinion about individual males' worth, and from the sires that passed his scrutiny he bred his superior animals. Any male creature, regardless of its heredity, that could consistently produce what Bakewell desired would be defined as an improved animal.[14] The genius of Bakewell's system, regardless of the gendered way it was structured, lay in the progeny test. But his major contribution to animal breeding would not be widely valued until the second half of the twentieth century.

Bakewell did not use a public registry system for pedigrees. His breeding records were personal, and therefore private. Public record keeping did exist for one animal breed by the end of Bakewell's lifetime, and some scholars believe the newly evolving system influenced him. By the late eighteenth century a public recording system had been established for the Thoroughbred horse in Britain.[15] Bakewell likely noted the increased selective breeding practices used in horse breeding rather than the recording system for them, because the Thoroughbred studbook was not initiated for breeding reasons. It arose out of the need to regulate horse racing.

Horse breeding had always commanded more interest than the breeding of farm animals because of the demand for specific types by the gentry, aristocracy, and royalty. This held true in both Europe and Britain. In Britain, horses used for war and for the pleasure of the nobles were generally chosen for size and strength until the end of the sixteenth century. During the seventeenth century, an increased emphasis on the navy for defense of an island nation resulted in a new demand for lighter and faster types, not destined for the cavalry.[16] Over the late seventeenth and early eighteenth centuries this trend culminated in the creation of the Thoroughbred, a superb racehorse. While selection took place on native horses, the development of the Thoroughbred owed much to the introduction of Arabian horse blood. In 1616 King James I brought to Britain the first Arabian, the Markham Arabian.[17] Because these horses had such desirable characteristics, more would be imported over the next two centuries to upgrade the evolving Thoroughbred. Intense selective breeding had already been practiced on this non-European animal, which encouraged the animals to produce true to type. Scholars have noted, however, that horses in hot environments like Arabia would be quite different from horses indigenous to the colder climate of northern Europe.[18] Regardless of human selection, then, Arabians naturally bred true to a type lighter, faster, and smaller than their European counterparts. Many Arabians would go into the makeup of the Thoroughbred, but three stallions in particular stand out: the Darley Arabian, the Godolphin Arabian, and the Byerly Turk (Arabian, in spite of the name).[19]

Thoroughbreds served the sport of flat horse racing, and it was that sport, not the infusion of Arabian blood, that triggered the need for public records. In 1791 falsified data in entries for races—such as age of the horse—led James Weatherby, with the blessing of the Jockey Club, to compile public records that would identify Thoroughbreds, and Arabians used in the breeding of Thoroughbreds, in the General Stud Book (GSB). (The Weatherby family, after James's death, continued to own and run the GSB for many years. Note that the first public studbook belonged to an individual, not to the breeders as a group.) Weatherby used available pedigree information, which described individual horses' ancestry, to help in identification. Quite simply, early pedigrees provided tools for documenting individuals' identity. The description of the Godolphin Arabian, published in volume 1, illustrates the importance of the Arabian to Thoroughbred breeding and shows how early record keeping for identification could be done when no information on ancestry existed. The record, or "pedigree," does not mention any sire or dam, nor could it be used for breeding purposes in the way a modern pedigree could.

The GODOLPHIN ARABIAN was a brown bay, about fifteen hands high, with some white on the off heel behind: there is a picture of him and his favorite Cat, in the library at Gog Magog, in Cambridgeshire, at which place he died, in the possession of [Lord] Godolphin, in 1753, being then supposed to be in his twenty-ninth year. That he was a genuine Arabian, his excellence as a Stallion is deemed a sufficient proof. In 1731, then the property of Mr. Coke, he was a Teaser to Hobgoblin, who refused to cover Roxanna, she was put to the Arabian, and from the cover produced *Lath,* the first of his get. It is remarkable that there is not a *superior* horse now on the Turf, without a cross of the Godolphin Arabian, neither has there been for several years past. Mr. Coke is said to have imported the above Arabian from France, and the Editor was once informed by a French gentleman, whom he has not had an opportunity of seeing since, that this horse had actually drawn a cart in the streets of Paris.[20]

The studbook clarified to everyone exactly which horse qualified for which race, and it also raised the value of animals by protecting potential buyers.[21] The earliest public studbook, then, had market implications. Just as the first component of this purebred system (the establishment of a breeding method with set rules) increased the animals' worth, so (apparently independently) did the second aspect of the system: public record keeping.

When George Coates began in 1822 to publish what would be the second public studbook in existence (for Shorthorn cattle) the new registry also used ancestry information, if possible, as a tool for identification. Shorthorn pedigrees in 1822 could be short because the ancestry of many foundation animals was unknown, but Coates was able to pin down which animal was which, just as the GSB managed to do for horses. That identification through public pedigree information was available for Shorthorns earlier than for other cattle breeds helped provide an important start to the breed's ultimate popularity and geographic expansion. Shorthorns established themselves as the most widely known breed of cattle in both Britain and North America until well into the twentieth century. Possibly Shorthorns became so popular, and played such a large role in the international cattle trade, not because they were improved before other breeds such as Herefords or Angus, but rather because of the head start provided by the breed's public herd book. Breeders of other breeds, who also relied on inbreeding principles, quickly saw that a public pedigree registry had helped to popularize, and thus sell, Shorthorns. Although Hereford cattle had a public pedigree registry by 1846 and a book for Angus cattle ap-

Eclipse, a famous eighteenth-century Thoroughbred. Eclipse won every race he ran and was known for his stamina as much as his speed. He sired some 344 winners over twenty-three years. Many people attributed the horse's greatness particularly to his Darley Arabian background on his sire's side. It is interesting, therefore, that through his dam the horse was more closely related to the Godolphin Arabian. *Eclipse*, by George Stubbs, 1770, oil on canvas, 39 by 50 inches. The Jockey Club, Newmarket.

peared in 1862, the delay seemed to have lasting effects on the market success of these breeds in relation to that of the Shorthorn.[22] Shorthorn breeders in Britain recognized the boost that the early existence of a herd book had given to the financial standing of the Shorthorn over the years and to the breed's advantage over others. "The registration of Shorthorn pedigrees is recognized as an inestimable boon to Shorthorn interest," one British expert wrote in 1876.[23]

From the beginning, pedigrees served an important function unrelated to hereditary information. That public records simply pinned down which individual was which (and therefore guaranteed proper identification) lay at the heart of the matter, and it would play an enormous role in the trade of stock that occurred over long distances. As North Americans entered the market for improved breeds of farm animals, buyers demanded publicly recorded pedigrees and set up public herd books in the New World to register their purchases. Over the rest of the nineteenth century, public herd books for other farm animals came into existence in North America and Britain for most of the improved breeds. Breeders in Europe found that they too needed public registry systems in order to sell stock in North America, and late in the nineteenth century herd books developed there. The first public herd books for cattle outside Britain appeared in France in 1855, in Germany in 1864, in Holland in 1874, and in Denmark in 1881.[24] In many cases public herd books existed in North America before they came into existence in the breed's country of origin, thus triggering European registry systems.[25] The history of Percheron horse breeding is a good example.

The importance of the American market to the industry cannot be exaggerated in this case, because before the First World War the market for Percherons existed only in the United States and Canada. At one point the largest herd of Percherons in the world lived on the Canadian Bar U Ranch. When the Percheron market spread beyond the New World, North America remained essential. Percherons began to enter Britain after the First World War, but they tended to do so from North America, not France. Because North America was critical for Percheron breeders in France, the American market drew them into using a public recording system. After Americans began to import draft horses from France, in 1876 they set up the National Association of Importers and Breeders of Norman Horses and a registry for Percherons. Canadians usually imported their Percherons from the United States, and for many years they used the American registry to pedigree their animals. In 1907 breeders in Canada established the Canadian Percheron Horse Association to look out for Canadian breeders' interests, and they began to register horses in a public herd book with the Canadian Livestock Records Corporation.

It became apparent shortly after 1876 that there were other varieties of draft horses in France and that no effort had been made to distinguish which ones were not Percherons. Everyone knew that a type, or even breed, of horse called a Percheron existed. But while selective breeding might have been practiced on the horses, the breeders in La Perche did not have a public registry, so American buyers found it hard to label individual horses as being distinctly "Percherons." Americans managed to force French breeders to establish public record keeping in 1883 and to accept for recording only draft horses born in the six provinces that had made up the old district of La Perche. American breeders would subsequently register imported horses as Percherons only if the animals had been entered in the French book.[26]

The new purebred breeding system resulted when public record keeping was linked to certain selective breeding practices ascribed to Bakewell. Public pedigrees could be used to certify breeding practices, however, and that connection proved equally significant in the long run. It introduced new notions about the word "breed" and even the meaning of breeding. Individual members of a breed could be described as "pure" or "impure," first based on pedigree or the lack of it and later based on different pedigrees. It quickly became apparent that pedigrees could be compared. Public registries that contained pedigrees worked meaningfully with an inbreeding system, because such pedigrees explained inbreeding levels, which in turn could be defined as purity to breed type in individuals. Breeders could also manipulate the system by maintaining or increasing certain levels of inbreeding and therefore could enhance purity. The less outcrossing, the greater the line's ability to breed truly, which came to be seen as an even better measure of purity to breed type. By 1870, when pedigrees could be related simultaneously to both purity and inbreeding, the philosophy behind this purebred system had achieved its mature shape. Animals were believed to be "pure" to breed type—therefore "purebred"—when they carried public pedigrees. But "purity" levels of such animals within breed type varied with the quality of their pedigrees. Ideas about breed, the meaning of purity within breed, and the role of pedigrees in breeding became entangled in a complicated way. Purity, pedigrees, and the related issue of quality ultimately made it necessary for breeders to set standards for entry in public records. What animals within any breed, breeders would be forced to decide, qualified for pedigree status? Of course purity, quality of pedigrees, and level of inbreeding quickly became marketing tools, and they affected the way the market worked as well.

An example of the conflicts over pedigree qualification and the market's role in pedigree regulation can be seen in a debate that arose in the Thoroughbred-breeding world. In 1913 Lord Jersey managed, by an act of Parliament, to bar Thoroughbred horses with any ancestry not registered in the GSB from entry in that studbook for breeding purposes. Some excellent Thoroughbred racehorses whose hereditary record could be found only in the American or French studbook could not receive pedigree status in the GSB and therefore could not function within the British Thoroughbred world. For over thirty years the Jersey Act barred some world-class racehorses from the GSB, claiming that the animals could not be proved to be "pure." American and French Thoroughbred breeders saw the act as having commercial motives, protecting British breeders from foreign horses.[27] The issue might have been touted as solely one of purity, but the clear market implications suggested otherwise. Utility—the ability to run—finally won out over purity in pedigrees as the prime qualification for breeding, however, and the Jersey Act was abrogated. Thoroughbred pedigrees could no longer act as tariff barriers under the pretense that "purity," and therefore quality, was the issue.

Purebred breeding created highly salable animals, and in the nineteenth century the method used to produce such stock was touted as a science. The system of inbreeding, it had always been believed, reflected a scientific method. The ability to control inbreeding by pedigree gave a veneer of scientific thinking to pedigree keeping as well and helped bind it intimately to a known method of practical breeding. The purebred breeding system came to be seen particularly as an agricultural science. In Canada, for example, examinations in agriculture at the newly established Ontario Agricultural College in the middle 1870s relied on a thorough understanding of Shorthorn history, the methods used in breeding Shorthorns from the end of the eighteenth century until the 1870s, and knowledge of the pedigrees recorded for those Shorthorns.[28]

While the fixing of type in farm animals by specific inbreeding practices had taught people that certain breeding methods seemed to control genetic traits and therefore explained the process of reproduction, some practical breeders argued that such issues were not well understood. While they considered their work scientific, they had difficulty defining exactly what was "scientific" about it. As early as 1876 an American Shorthorn breeder stated that science had in fact done little to help in breeding better stock.[29] In 1896 a Canadian farm journal reported in an article titled "The Science of Farming" that "in spite of the great advances that have been made by breeders of late years in the science of breeding, there are still many things either but little understood by them or totally beyond their comprehension." The journal added, "The influence of the male on his offspring is evidently . . . a moveable quantity. This rule holds good, too, in the human race. One who investigated the subject found that the firstborn resembles the father the most, and this is especially noticeable when there is a great affection existing between the parents."[30] The writer apparently believed that people did not understand how heredity worked. The quotation also reveals sustained allegiance to fundamental principles of the Bakewellian system: that males represented the most significant element in heredity and that breeding for improvement could be done only on male lines. The writer also seems to have believed that the breeding of people could be related to the breeding of animals. Interest in human genealogy had in fact dovetailed with concern over animal pedigrees in the nineteenth century. *Burke's Peerage* first appeared in print in 1826, and John Burke published his work on genealogy in 1833. Scholars have noted that human "pedigrees" and the tracing of people's generational backgrounds became common about the same time that the earlier public studbooks and herd books emerged.[31]

By the twentieth century, purebred animal breeders hailed the new science of genetics as their salvation. In 1901 Gregor Mendel's laws, established in 1865 from the monk's work on peas, were rediscovered. In the United States it would be agricultural organizations, not medical institutions or biology departments of universities that first explored the implications of these rediscovered ideas. In 1903, for example, members of the American Association of Agricultural Colleges and Experimental Stations founded the American Breeders' Association with the intent of using Mendelian laws to raise productivity in agriculture and also to provide for research in the new field of genetics. The organization began to publish the *American Breeders' Magazine*, which changed its name to the *Journal of Heredity* in 1913. Research in genetics within the United States initially concentrated on the breeding of plants and animals, not of humans.[32]

Information derived from research on plant and animal breeding could be applied to people, though, and even early statements in the journal published by the American Breeders' Association blurred the organization's focus—concern with animal or human genetics. In 1910 the journal said:

> The objects of the American Breeders' Association are the advancement of the discovery of the basic facts concerning heredity, the devising of new plans for creative breeding, and the organization of those projects which led toward improved plants, animals, and men. It functions on the skirmishing line to discover the lay of new land, helps to plan the campaign and to direct advances where the largest results may be secured. It is also alive to the work of the great army of active practical breeders which must carry out the bulk of improvements. . . . The Association promotes and helps organize cooperation among those interested in studying the laws of heredity, those devising improved methods of breeding the new respective species of plants and animals . . . those who multiply purebred animals . . . and those interested in eugenics or heredity in man.[33]

Initially, agriculturalists across North America believed that genetics would provide advances in livestock farming. "The breeder's art has been profoundly modified by discoveries made within the last decade, yet we are only at the beginning of a new era in which a knowledge of the principles involved gives us a science of breeding," the American Breeders' Association stated in 1910.[34] One Canadian farm journal had the following to say in 1907.

> Recent discoveries with respect to the laws [of heredity] have furnished the breeder a flood of light on the subject of [breeding], in comparison with which [the breeder] may have been said to have formerly been groping in the darkness.... The chief of these discoveries or rather rediscoveries (for it is a remarkable fact that these discoveries were first made some forty-five years ago, but were immediately lost sight [of] and remained buried in oblivion until recently discovered within the last five years) is the fact that the seed which an animal (also a plant) secretes for the reproduction of its kind is not all alike, but that some of its seed will produce certain qualities and the seed of the same animal will produce other qualities, sometimes radically different or opposite qualities, so that the same animal may secrete six, eight, ten, twelve, twenty and many more varieties of seed.[35]

It became apparent not very long after the founding of the American Breeders' Association, however, that while genetics could influence plant production, the science provided no useful tool for animal breeding. In 1912 the organization stated that the purebred system remained the only way to improve animals. "A science of animal breeding scarcely yet exists, not because thoughtful men have failed to give attention to the subject but rather because of its inherent difficulties. We breed animals as our fathers and grandfathers [did] because their time-honored methods succeed and we know of no methods of changing these methods. Indeed we can not expect to improve them in a rational way until we learn why certain methods succeed and why others fail."[36] By 1918 most concurred that genetics had contributed little "toward improvement of the existing methods of animal breeding."[37] And in 1925 a farm expert wrote: "Up to the present time, the new knowledge of genetics has contributed little" to advances in animal breeding.[38]

Even by the end of the twentieth century some still believed that the pure science of genetics had done little to help in the breeding of farm animals. "Our knowledge of the genetic architecture is limited [even] in laboratory animals . . . and is very limited for economic traits of farm animals," one expert stated in 1998.[39] Mendelian genetics might show some signs of influencing breeding, but for the moment the new breakthroughs in understanding DNA, and in the ability to clone genes, can affect only small numbers of laboratory animals—even if these animals happen to be farm mammals—not large herds and flocks or pet populations. However, this situation might change quickly. For example, at the time I wrote this book, a two-year-old cloned Holstein named Starbuck II, ready for service, had the capacity to affect hundreds of thousands of cows. Starbuck II was cloned from Hanover Hill Starbuck from cells taken from that bull just before he died in 1998. The original bull, born in 1979 in Canada, was credited with "revolution[izing] Holstein breeding worldwide" and left 200,000 daughters.[40]

The spread of the eugenics movement in the early twentieth century only complicated the relationship between purebred breeding and the science of genetics. In 1883 Francis Galton (born in 1822 and a cousin of Charles Darwin) coined the word "eugenics." Controlling the racial qualities of future generations of people concerned Galton, and he hoped to find a way to weed out the physically and mentally unfit members of the human population. Darwin's theory of evolution and survival of the fittest, Victorian concern for the decadence of the lower classes, and the Malthusian threat of overpopulation all influenced him, as did selective animal breeding and its recording in public pedigrees. Galton believed that a pedigree system could be used to improve humans. Although concern with genealogy and with the family background of the elite classes existed by Galton's time, his sense that human pedigrees could be used scientifically was new. The study of human pedigrees, his followers in Britain and North America would argue, could determine which people should be labeled as unfit to breed.[41]

Between 1891 and 1906, Karl Pearson, professor of applied mathematics and mechanics at University College in London, developed a system of statistics called biometry, which could be applied to biology and therefore used by eugenicists. Both American and British eugenicists proceeded to use biometry to analyze human pedigrees. British eugenicists did not see their results in Mendelian terms, but their American counterparts did. Americans, such as Charles Davenport, hoped to combine the implications of Mendelian genetics with eugenic aims through a synthesis of pedigree analysis.[42] In the United States, eugenicists argued that their work in biometry would explain laws of heredity. Their counterparts in

Britain believed that pedigrees and their interpretation revealed the results of heredity but did not explain it.[43] A cleavage over the value of biometry developed between geneticists and eugenicists. Geneticists who pursued Mendelian genetics refused to accept biometry, arguing that the system did not show how heredity worked.[44] Regardless of the controversy around biometry, the system did reintroduce the Bakewellian principle of progeny testing as a method of improvement through breeding. At this point the progeny test did not affect animal breeding, but it became entangled with attitudes toward human breeding.

Geneticists and eugenicists might not have agreed over the values of biometry, but clear demarcation between the two groups did not always exist in the early twentieth century. Eugenic thinking increasingly infiltrated genetic thinking. The disenchantment of animal breeders with genetics early in the twentieth century escalated with its greater connection to eugenics. Purebred breeders might have noted similarities between human and animal heredity patterns, but they disdained strident eugenic views. A review of early twentieth-century Canadian farm journals shows clearly that breeders resented having their work directly linked to pure eugenic theory, even though eugenicists openly touted the idea that their own work built on that of purebred breeders.[45] The early eugenic component of organizations devoted to genetics (by 1920 the American Breeders' Association had virtually been taken over by eugenicists) made it hard for breeders to see genetics outside eugenics. The failure of genetics to help them only weakened any propensity for them to support either movement. As a result, they tended to further emphasize the artistic aspect of their work. They also became less likely to see purebred breeding as "scientific," or even as innovative in a scientific sense. In 1925 a farm expert wrote, "Animal breeding proceeds in much the same way as it [did] four thousand years ago."[46]

Breeders had always seen their work as creative, and they often explained the success of certain breeders in terms of their artistic vision.[47] Aristocratic breeders in Britain collected paintings of their improved stock. Over the mid-nineteenth century the 3d Earl of Spencer commissioned a large number of paintings portraying his Shorthorns. In Britain paintings of famous animals began to be reproduced cheaply in vast numbers and sold profitably. As the nineteenth century advanced, North American farm journals increasingly carried lithographs of paintings of improved livestock. As one Canadian farm journal stated in 1880, "The language of the eyes is the only universal language."[48] Artistic images could even be viewed as useful tools for a breeder.[49] By the early twentieth century, artistic imagery and discussion of the purebred breeder's "art" became even more prominent. In the journal of the American Breeders' Association, readers would be more likely to find purebred breeding head-lined as the art than as the science of breeding, which had been more common over the nineteenth century. Purebred breeding was, one might say, the art of genetics.

While art continued to play a substantial role in purebred breeding, after the mid-twentieth century the system would again rely on science and be seen as a scientific as well as an artistic method. By then genetics provided useful guidance in selecting individual animals for breeding. Mendelism was less important to that pattern than the study of population genetics. The Bakewellian idea of progeny testing survived the decline of eugenics, which rapidly spiraled downward because of events in Hitler's Germany. Statistical analysis of whole populations (population genetics) became part of genetic research when it became concerned with heredity as expressed in populations, not individuals. Some geneticists practicing in this field turned their attention to farm animal reproduction. Scientists such as Alan Robertson in Britain began exploring how quantitative work on farm animal populations could explain genetic characteristics as expressed in populations. Robertson concentrated much of his research on improvement in dairy cattle. He studied progeny testing in these animals by looking at the productivity of bulls over a fifteen-year period. Technology made the work easier than it had been in Bakewell's time. Artificial insemination allowed for the extensive breeding of one bull over various cow herds. Robertson found that certain bulls could produce desirable cows while others proved better at siring desirable bulls. He revealed that it was simplistic to say that a bull produced good milking daughters. Robertson, in effect, fine-tuned Bakewell's progeny testing system.[50]

It has been increasingly fine-tuned since the 1960s and used to evaluate purebred herds and individuals within them. Data on the individual's productivity compared with that of its peers (for example, milk production) often became part of a pedigree. The science of genetics has also been able to locate certain genetic problems, such as hip problems and Collie eye anomaly in dog breeds and has allowed breeders to avoid them. In more modern times genetic testing results might also be added to pedigrees to show freedom from hereditary defects.

In the late eighteenth and early nineteenth centuries a breeding system took shape in Britain. The association of public ancestry record keeping with specific eighteenth-century breeding procedures established a purebred breeding system that produced valuable animals. Animals bred by this method became desirable in North America and thereby created an industry. A number of movements between the late eighteenth and early twentieth centuries helped shape the basic principles that constituted this breeding system: growing genetic understanding after 1900 in the North Atlantic triangle; attitudes toward the eugenics movement; the conviction that the breeding of good animals resulted partially from artistic skill; the acceptance of genetics as a science after midcentury; and most critical of all, a transatlantic trade. Purebred breeders supplied the market with a product made through the artistic manipulation of genetics. This breeding system created most of the numerous recognized breeds of animals that we know today. Each breed has its own history, which reveals how breeders regulated themselves, the standards they set for breed qualification, what they wanted to do over the years, and how they and their products fared in relation to other breeds.

Some premises that went into the evolution of purebred breeding have long since fallen into disrepute. But the system has endured. Perhaps one of its most surprising features is how little its practices have changed over the past century. Perhaps equally surprising, enormous advances in science generally and in genetics particularly over the past fifty years have hardly made a dent in its fundamental aspects, its perceived ability to create excellence, or the marketability of its products. Purebred breeding remains very much with us in the early years of the twenty-first century.

# Two

## SHORTHORNS AND ANIMAL IMPROVEMENT

"It is claimed ... that the Short-horn blood produces 'the farmer's cow' *par excellence* of the world," the American breeder Alvin Sanders wrote in his history of the Shorthorn.[1] Sanders referred to a breed of cattle that by 1860 had achieved great popularity in both Britain and North America. The fusion of breeding principles with public pedigree keeping occurred in conjunction with the creation and early evolution of the Shorthorn. Shorthorn history, then, does much to explain patterns of purebred breeding that came to characterize the system. This chapter explores the Shorthorn world from 1790 until just after the First World War by dealing with breeding practices applied to Shorthorns, how public registry keeping took shape within that environment, government influence on the Shorthorn industry, and the way each interacted with market conditions.

Shorthorns evolved as a breed over the late eighteenth and nineteenth centuries in Britain, a country where farmers favored cattle husbandry. Conditions in Britain at that time encouraged the production of all livestock. Increasing volatility in wheat prices—a result both of wars and of poor British harvests between 1780 and 1815—made wheat cultivation (arable farming) less attractive and livestock farming (pastoral farming) more attractive to farmers and the nobility alike. The instability continued in peacetime, with low prices prevailing for the better harvests after 1821. A series of tariffs known as the Corn Laws were passed to help stabilize the situation, but the trade in wheat remained volatile. The landed classes increasingly found that their incomes fluctuated when they relied on wheat cultivation, and many became more committed to livestock. Parliament abolished protection of wheat at midcentury, and that action did even more to stimulate livestock production. Research suggests that Robert Peel supported the repeal of the Corn Laws in 1846 in order to promote the livestock interests of the nobles, not to provide cheap food for the urban poor.[2] A period of great agricultural expansion took place on British farms after the repeal and lasted until the recession of 1873.

Within this age of "high farming," livestock production was increasingly important. Even though the agricultural markets appeared to remain depressed until general economic recovery in 1896, livestock farming, and especially the production of improved farm animals, continued to be profitable.[3] Over the second half of the nineteenth century Britain witnessed a sustained concern for better livestock farming, even more so than earlier in the century or in the eighteenth century.

This rising interest in livestock husbandry encouraged British cattle breeders to work with a variety of bovine types, trying to improve beefing qualities in particular. While there was feverish activity in the buying and selling of Bakewell-bred Longhorns at the end of the eighteenth century, agriculturalists also attempted, through inbreeding, to enhance the beefing qualities of other sorts of cattle. Such strategies led to better standardization of cattle into "breeds" and also to the geographic expansion of breed types. The Devon variety had been concentrated mostly in the West Country, but as farmers recognized its beefing potential the animals soon became more widespread. The Hereford breed slowly expanded from where it had originated as agriculturalists experimented with it. The Sussex, very like the Hereford and Devon, for some reason remained unknown outside that county. Angus and Galloways, while still centered in Scotland, had started to move into Midlands England, even if only for feeding. The varieties that would ultimately become the Shorthorn were also experimented with for beefing, and they began to spread outside their original locale—in this case Lincolnshire and Yorkshire—more quickly and in greater numbers than any of the breeds noted above.[4]

Charles and Robert Colling in particular worked with these kinds of cattle found in the Midlands and northern England, and the brothers receive the credit for creating the modern Shorthorn. The cattle types they worked with had been named for the regions the stock came from: Durham, Teeswater, and Holderness being three examples. While these early animals could be described as variable in type, generally speaking they tended to be cattle of great size, good milking ability, and poor beefing qualities. About 1780 the Colling brothers began to practice intense inbreeding on the stock to enhance beefing characteristics. The brothers did not always acquire foundation animals based on hereditary background. To them, individual worth counted more. They were even prepared to use cattle outside Shorthorn varieties in breeding programs designed to create a new, uniform Shorthorn; for example, they introduced Galloway blood. Certain specific individual animals the Collings worked with would be of lasting importance to the subsequent fortunes of the Shorthorn. In

1784 Charles Colling bought a cow named Duchess from a tenant farmer, and she remained, in his opinion, the best cow he ever saw. He hoped to perpetuate her, not improve on her. Meanwhile Robert Colling spied a good bull grazing by the roadside. Colling bought this animal, named Hubback, from a local bricklayer, even though the bull's ancestry was unknown. A particularly good grandson of Hubback, Foljambe, impressed Charles Colling. He decided to breed a daughter of Foljambe (named Phoenix) to a son of Foljambe (named Bolingbroke), and that mating produced the bull Favorite in 1793. While Favorite reflected intense inbreeding to Foljambe, he showed obvious inbreeding to Hubback as well. (Favorite, inbred to a degree of 19.2 percent, would be bred back to his daughters in some cases up to five or six generations, thus maintaining an unnaturally close breeding index in later generations to both Favorite and Hubback.)[5] One of the old Duchess cow's descendants (Young Duchess), sired by Comet (a bull heavily inbred to Favorite on his dam's side as well as sired by him), would herself go on to found a new line of inbred stock—largely from an intense program of inbreeding to Favorite.

The Colling brothers publicized their breeding by linking it to Bakewell's methods, and their efforts at promotion at least partially explain why the new Shorthorn began to catch the attention of both the nobility and the general agricultural world. Shorthorns started to expand beyond their place of origin and soon could be found all over England, Wales, and southern Scotland. The cattle came to be seen as an innovative type of stock, capable of better meat production. Early appreciation of the Shorthorn, however, had nothing to do with publicly kept pedigrees. No public record-keeping system existed. That situation would change early in the nineteenth century, and when it did the fortunes of the Shorthorn improved further.

In 1800 Thomas Bates began buying stock from the Collings. At the brothers' dispersal sale in 1810, Bates acquired Young Duchess. Bates bought her on the strength of her hereditary background, not her looks or her own innate quality. Many at the time found her shabby in appearance, and Bates's own father ridiculed his son for the purchase.[6] Bates's action, choosing foundation stock by hereditary background, not individual worth, represented a departure from the principles the Collings had used. Bates renamed his new cow Duchess 1st and started on a program of intense inbreeding to her. In effect he practiced inbreeding to the Colling bull Favorite (through Comet) and followed the breeding strategies of the Collings. He also claimed that the Duchess line represented a family of immense purity, maintained by inbreeding. Over eight generations of cat-

Duchess by Daisy Bull. Thomas Bates purchased Young Duchess, great-grand-daughter of the Colling-bred cow Duchess by Daisy Bull, and in the end he produced from her a highly successful and heavily inbred line of Shorthorns. Note the stylized image of Duchess by Daisy Bull—thin legs supporting a blocky, heavy body with a very small head. Cattlemen from the late eighteenth century to beyond the mid-nineteenth considered these features marks of improved cattle. Lithograph by J. R. Page, from the 1874 *History of Short-Horn Cattle* by Lewis F. Allen, founder of the first public registry for Shorthorns in the United States.

tle, Bates produced from this Duchess family sixty-three cows named Duchess 2d to Duchess 64th and about forty-five males of the same line, including some twenty-nine Dukes. Bates maintained a level of 40 percent inbreeding (meaning the stock was 40 percent more inbred than the average Shorthorn, which would be more inbred than the average nonpure-bred cow).[7] The Duchess line had never been prolific, even under the Collings' breeding strategies, and Bates could produce on average fewer than two females a year within the Duchess line. Sometimes he held over cows that had not calved for several years. He refused, however, to use anything but Duchess-related bulls on his favorite family of cows. All others, he claimed, could be described as animals that were "impure." As one early Shorthorn expert stated, "He would not admit that other contemporary bloods were worthy of being crossed upon his Duchesses."[8]

When George Coates began in 1822 to publish a herd book for Shorthorns, Bates actively supported the new public registry system, and some scholars suggest that Coates began his work largely at Bates's instigation.[9] Publicly recorded pedigrees, under Bates, would become just as important as inbreeding in maintaining consistency of type. Bates also contended that different pedigrees demonstrated different degrees of purity. He argued that the pedigrees of his own stock represented the highest level of purity available for Shorthorns. After Coates died, Bates claimed maximum purity for his own breeding by stating that the herd book had not been exclusive enough, and he even described some of Charles Colling's renowned stock as "mongrel." He did not want his high-class Duchesses associated with this lack of purity.[10] His obsession with the idea that pedigrees revealed degrees of purity, which in turn certified quality, seems to have been driven more promotionally than theoretically. Everyone at the time knew that the genealogy of foundation animals, admired for their perceived quality and therefore used to fix a breed close to their type, was simply unknown.[11] The background of Colling's bull Hubback serves as an example.

While the general popularity of Shorthorns rose over the early nineteenth century, Bates's work encouraged more particularly the breeding of Shorthorns by the elite. Bates's public description of how he bred the Duchesses and his connection of pedigree to purity (and therefore both quality and value) increased fanciers' interest in Shorthorns. The breed had attracted "fancy" notice in the Collings' lifetime late in the eighteenth century and based on Collings stock early in the nineteenth. (The 3d Earl Spencer, for example, started in 1818 to build up a large herd of Collings-bred Shorthorns.)[12] But that side of Shorthorn breeding grew faster as a result of Bates's work. Three distinct types of owners—two groups of breeders (experts and wealthy amateurs) and investors—made up the rapidly evolving fancy section, and specialized markets arose to serve them. Shorthorns of Bates breeding were valued as precious jewels, and the animals were bought and sold with that market in mind.[13] New, elite markets shaped notions about the meaning of pedigrees in any breeding program. After the mid-nineteenth century, attitudes toward all pedigrees and their relation to any breeding would reflect Bates's principles as well. Publicly recorded pedigrees came to define quality through perceived purity. Bates's fancy breeding ideas, then, would ultimately affect everyone's perception about what it meant to define a cow as a purebred Shorthorn. Fancy breeding and its linkage to pedigree came through this connection to shape all aspects of the Shorthorn market, and conversely to be shaped by it.

This complex transformation in British livestock production had not gone unnoticed in North America. Agricultural practices there had changed profoundly, largely because of greater food requirements as the population expanded from both internal growth and immigration. Increased demand for meat drew attention in both the United States and Canada to British efforts at livestock improvement. Support for purebred animal breeding by the upper classes in Britain also commanded the interest of wealthier men in the New World. Because expansion occurred earlier in the United States, breeders there investigated Britain's cattle improvement techniques before their counterparts in Canada. By 1800 the rapidly growing population had spread well into the interior. Patterns of cattle farming and beef raising that were well established in Virginia, Maryland, and Pennsylvania before the American Revolution were extended into Ohio country early in the nineteenth century, and interest in improved stock grew. In 1817 a number of wealthy men in Kentucky, concerned for better cattle in the Midwest, decided to import English cattle. They selected both Longhorns and Shorthorns. The stock left a mark on the local cattle population, and its quality was recognized. The animals did not carry publicly recorded pedigrees, because in 1817 there was no public record-keeping system for cattle. They and their descendants became known as the Seventeens, and their fame quickly spread beyond Kentucky. Quality cattle that had descended from the Seventeens entered Ohio after 1820.

Cattle increasingly would constitute an important sector of the economy in the Old Northwest. As early as 1808, areas like the Scioto Valley were supplying meat to the eastern seaboard, and by 1830 Ohio had become the American beef cattle center. Some Ohio cattlemen, who knew about the excellent influence of the Seventeens, decided to import more of the improved cattle from British breeders.[14] One of the most enterprising was Felix Renick, whose family had settled in Ohio early in the nineteenth century. In 1833 Renick and a number of others formed the Ohio Importing Company for Importing English Cattle. Forty-eight people held ninety-two shares, worth $100 apiece, meaning the company raised $9,200 of capital. Renick led a delegation that made several trips to England. The shareholders had agreed that no particular breed should be sought and that pedigrees would not influence selection. They wanted only quality. In 1834 Renick and his group bought nineteen head of cattle, in 1835 they bought seven more, and in 1836 the delegation under Renick purchased thirty-five animals.[15] Early in this buying work, the Americans met Bates at the Darlington market.

Bates courted the Americans, but Renick refused to be drawn in too quickly, and the company had agreed it would not necessarily buy Shorthorns. Renick did not believe that pedigrees dictated quality—he had seen too many good cattle that were descended from the Seventeens.[16] The effects of excessive inbreeding also concerned him. But Renick quickly decided that although pedigrees did not indicate quality, they did raise an animal's monetary value. He learned more about public pedigree keeping, and he studied Coates's herd book. He wrote the following to shareholders in Ohio: "The value depends almost entirely upon the purity of the blood and high pedigree." He continued: "Thus you see the situation we are placed in. We must take either cattle without pedigree or much of anything else to recommend them or take those that have at least pedigrees, with more excellence of form and size, at a higher price. The latter was in our judgment the better of the two alternatives and the one we have so far pursued, and shall continue to pursue, and take fewer in number. . . . We want none without fair pedigrees, but form and size they must have or they will not be well received here."[17] Renick had decided that to protect the shareholders' investments he should take cattle that had pedigrees under a public recording system. It seemed to him that if two animals were of equal quality, the pedigreed one could command more money and would maintain its worth better. It was simply good business to invest in stock registered in a public book. In 1833 only the Shorthorn had a public herd book, so Renick bought Shorthorns.

He also purchased Bates cattle (a decision with lasting significance for American Shorthorns until the end of the century), and even a few Duchesses. Bates understood the importance of publicly recorded pedigrees in the transatlantic market, and he continued to correspond with the Ohio farmer. "My object has never been to make money by breeding," Bates wrote in 1837, "but to improve the breed of Short-horns; and I know . . . I will not sell any to anyone who has not the same object in view. On this principle I began breeding, and I am convinced I have a better breed of Short-horns in my possession at present than there has been for the last fifty years, even in the best days of the Messrs. Colling."[18]

The value that pedigree keeping in public books held for transatlantic trading became evident in the United States when the Ohio Importing Company auctioned the stock for high prices. An elaborate catalog containing complete pedigrees of the animals offered accompanied the first sale, held in 1836. Nineteen bulls sold for an average of $790, and twenty-four females brought $800 on average. The 1837 dispersal sale of the last of the stock realized an average of $1,180 on six bulls, and nine females averaged $1,070.[19] These prices were remarkable when compared with typical yearly incomes. Several hundred dollars a year would have been a good amount in Ohio country at that time. Proceeds from these sales generated about $50,000, a good profit from the shareholders' $9,200 investment. But the investors made an even better profit than that, because they kept some of the stock they bought (or calves born from those animals). Apparently buying pedigree did pay. The influence of these Shorthorns was felt throughout the Ohio Valley as their genetics spread into local cattle. While the reputation of the newly acquired Shorthorns rose, that of the Seventeens fell. Ironically, the cattle that had stimulated the formation of the Ohio Importing Company lost their exalted position when the newcomers arrived, simply because the Seventeens did not have recorded pedigrees. Bates cattle and recorded pedigrees increased in value over the next decades as breeders in both Britain and the United States attempted to preserve the "purity" of the lines by more intense inbreeding. Bates's promotional ideas stressing inbreeding, which could be verified in pedigrees, influenced cattlemen who were not simply interested in fancy cattle. Even Renick began to inbreed the Duchess stock that he had kept after the sales.[20]

Interest in good beefing cattle generally and Shorthorns particularly—Bates or otherwise—continued to climb as the cattle centers of the Midwest expanded. By the 1850s cattle-raising had spread to Illinois and Iowa. Railway building coincided with the expansion of corn growing and cattle fattening into all of the Old Northwest. Trains allowed poorer stock to move farther to markets, and by 1870 stockmen in the old cattle grounds of the Midwest found they could compete with the more distant and cheaper western range and Texas stock by breeding better livestock—animals that could provide more meat in a shorter time. Kentucky cattlemen in particular developed a strong interest in Bates genetics and especially in Duchesses. These breeders joined others in Britain and in central Canada in the pursuit of this line of Shorthorn. Fancy and utility became blended in their minds, and they took Bates's promotional work as instruction on how to breed the stock. After Bates died, breeders in both Britain and the United States began to inbreed in particular to his already inbred Duchess cattle, and the inbreeding level of the Duchesses rose above 40 percent. One bull bred in Britain in the 1860s, for example, was reputed to be inbred on Bates lines for over twenty generations.[21]

Inbreeding and purity began to command the interest of rich fanciers in New England and New York. These wealthy men acquired Bates Duchess cattle to some degree for show ring use, but more for investment. One important Shorthorn owner from New York, Samuel Thorne (a wealthy man in the leather business in New York City), began to collect stock by importing Bates cattle. In 1861 he visited England and found

English breeders interested in his Duchesses. By that time a family known as the Grand Duchesses represented the only surviving line of pure Duchesses in Britain, and over the 1860s their number declined owing to natural infertility in all hereditary Duchess stock. Rarity raised the demand for Thorne's Duchesses in both Britain and the United States. In 1866 Thorne held a dispersal sale, and J. O. Sheldon, a rich New Yorker, bought the entire herd. One contemporary Shorthorn expert described the Duchess market in 1866 as follows. "This gave [Sheldon] a monopoly on so-called 'pure' Duchess blood in America; and as the English landed proprietors, as well as prominent Kentucky breeders, were developing a marked preference for [Duchess blood] he now occupied a strong speculative position." In 1867 Sheldon exported Duchesses to Britain.[22] By the late 1860s a complicated Shorthorn market had developed between Britain and the United States, and certain breeding methods had evolved that relied on both public pedigree recording and inbreeding. Within that environment, Canada played a critical role. By that time, powerful interests in livestock production and purebred breeding existed there.

When the Canadian nation came into existence with Confederation in 1867, agriculture in Quebec and Ontario, the most populated parts of the country, had been transformed, and farmers there had entered a new international market for farm products.[23] The United States had dissolved into Civil War in 1861, and Canadian farmers quickly found opportunities beyond the traditional exporting of wheat. The war cut off the western cattle-producing sections of the United States from the eastern consuming states, causing a shortage of beef.[24] This situation gave an impetus to cattle production in central Canada and triggered a lucrative international trade in livestock, which survived the abrogation of the Reciprocity Treaty in 1866. In 1870, for example, Ontario alone exported nearly 100,000 head of cattle to the United States.[25]

A healthy foreign market for cattle from central Canada supported the growth of beef cattle raising, and that in turn stimulated an interest in the purebred beef breeds in Britain. While by the 1850s Canadians (mostly English, not French, Canadians) began importing purebred beef cattle from Britain, Canada's purebred industry would not really get under way until the late 1860s. Generally, wealthy men pursued purebred livestock breeding, for complex reasons relating both to hobbyism and to nationalism. These men also tended to be importers of purebred livestock rather than breeders.[26] Not being experts in cattle breeding, they relied (as did wealthy American buyers) on the advice of managers who were often British-born immigrants to the New World. And several of these individuals—importers and managers—became part of the rapidly developing international Duchess Shorthorn situation. Mathew Cochrane, a Canadian senator from the Eastern Townships of Quebec, began buying Bates Duchess cattle in England in 1868. Simon Beattie, a Scottish immigrant who married into a prominent Ontario Shorthorn breeding family, became Cochrane's manager. Beattie worked closely with John Hope, another British immigrant who favored Bates cattle and who served as the manager of Bow Park, the estate of George Brown, Ontario father of Confederation.[27] As one contemporary expert said, Cochrane "resolved to be 'in' on the Duchess proposition." He took Beattie's advice on what to purchase in England. Cochrane bought from Sheldon as well, and he could pay very high prices. His move into the market at this particular time drove up prices on a scarce commodity just when several wealthy men in New York started to amass herds of pure Duchesses.

By 1870 Samuel Campbell of New York State had acquired the largest herd of Duchess stock, purchased on the advice of Richard Gibson, who would be a significant figure in the purebred world in all three countries of the North Atlantic triangle. Gibson, who was born in 1840 in Britain and in 1861 had immigrated to Canada, loved the tall, beautiful Bates cattle of his native England, and he persuaded Campbell to buy Bates Shorthorns. In 1869 Gibson went to England to import Duchess foundation stock for Campbell's new herd in New York Mills, New York. In 1870, on Gibson's advice, Campbell bought Sheldon's Geneva herd. By 1873 Gibson decided that the speculation in Duchess cattle had reached a peak, and he advised Campbell to sell out. At the New York Mills sale, buyers from the United States and Britain (only Simon Beattie came from Canada) converged for the sale of the world's last pure Duchesses. One in particular, 8th Duchess of Geneva, was descended from Duchess 55th, an animal that had sold at Bates's dispersal sale in 1850 for the sterling equivalent of roughly U.S. $500.[28] When the seven-year-old red-and-white cow left the sale ring, she had been sold to an Englishman for the staggering sum of $40,600, at a time when yearly incomes in North America averaged less than $500. Within a few days the cow delivered a stillborn heifer, and she herself died shortly thereafter. One observer at the sale wrote, "One long breath, and then the cheers went up, and thousands there seemed fairly beside themselves, and extravagant things which were said and done would fill a volume."[29]

Surprisingly, the market for Duchess cattle did not break with the New York Mills sale. In fact, it did not collapse until the loans procured to buy the expensive cattle began to come due. Cochrane, for example, had a very successful sale of Duchess cattle in 1877 in Britain. He held a final sale in 1882 in Chicago, where the chief buyer was Bow Park, the farm of the late George Brown. Richard Gibson, who had returned to Canada after 1873 and bought a farm near London, Ontario, also sold out his Duchesses at Chicago in 1882. Kentucky breeders continued to buy the stock as long as they could. But infertility finally took the ultimate toll. By 1883, regardless of whether a market existed for them, none of the cattle remained. Pure Duchesses had become extinct.[30]

The Duchess speculation appalled Shorthorn breeders generally because it clearly proved that the market could be the major force in breeding decisions and because the boom was utterly unrelated to livestock improvement. In 1876 a British Shorthorn expert wrote that "American prices gave an impetus to both business and breeding here," but he claimed that had simply become "a fashion and a fancy."[31] In the same year, American Shorthorn breeders held a meeting in Toronto, and the issue stimulated hot debate. One Ontario breeder stated: "I think the high breeding has done a great deal of harm in this country, has driven the small breeders out of the country, and has confined the breeding to very high and aristocratic breeders."[32] Judge T. C. Jones spoke at length about Shorthorn breeding, the relation of pedigree keeping to good cattle, and the influence of inbreeding on soundness. One should seek not the perfect pedigree, he said, but the perfect animal. However, he clearly admired pedigrees as a breeding tool. Pedigrees, he pointed out, supply information on what stock should be used for inbreeding. He bemoaned the influence of money on the breeding of functional cattle. Pretty animals might not be sound, he noted, but they will command a lot of money, and that stimulates the breeder to breed more such stock.[33] Shorthorn breeders did not forget the New York Mills sale. In 1897 the president of the Dominion Shorthorn Association (Canadian) called it "one of the worst days in Shorthorn history."[34] By the end of the century, breeders had come to view that day as a watershed between a time when pedigree, inbreeding, and market factors governed Shorthorn production and a time when the desire to breed for true "improvement" again, fortunately, held the upper hand.

In the early twentieth century Alvin Sanders, the great American cattle expert, wrote this about the Duchess crisis and its relation to utility and fancy.

The wealth of the cattle-breeding world was now ready to be poured at the feet of the Shorthorn. The beauty and practical utility of the breed captivated the great landed proprietors of both hemispheres, as well as the farmers and feeders of both continents; and under the stimulus of a demand almost world-wide in its character those who had the means to gratify their taste for rare specimens of the breed were forced to measure values not so much by the mere intrinsic worth of individual animals for the feed-lot or the dairy as by the degree of personal satisfaction flowing from the ownership of Shorthorns of illustrious lineage or bearing the badge of show-yard superiority.[35]

Long before the craze for Duchess cattle peaked, a number of farmers had begun breeding for "the better usefulness" in the Shorthorn that Jones described at the American Shorthorn Breeders' Association meeting. A new center of particular importance to this movement arose in Aberdeen, Scotland, where many chose not to rely on either pedigrees or excessive inbreeding. These breeders also dismissed both the show ring and speculation as arbiters of quality, and therefore of value. The elite section of the Shorthorn breed that had promoted fancy through overemphasis on both pedigrees and inbreeding seemed to them to have undermined the credibility of the Shorthorn with men who wanted to produce useful farm animals. Amos Cruickshank became the most significant of the breeders who responded to the need for good utility Shorthorns. Born in 1808, the son of a tenant farmer, in 1837 Cruickshank went into partnership with his brother Anthony and rented a two-hundred-acre farm named Sittyton. In their new Shorthorn breeding venture, Anthony supplied the finances and Amos the expertise. The brothers bought foundation stock for Sittyton in 1837 and paid a good deal for what they considered good cattle. From the beginning, Amos had contempt for pedigrees that went with animals lacking constitution, substance, and quality. As one Canadian breeder later said, "Mr. Cruickshank never followed fashion either in pedigree or any other point, but had his own sound common sense to guide him."[36] The brothers went to the Bates dispersal sale but did not buy much. Amos did not select fashionable cattle for his foundation herd, nor was he prepared to keep animals that would not produce what he wanted.[37] Cruickshank would be breeding cattle, based on his convictions, for forty years before his style became popular. As late as 1876 he noted that all Shorthorn sales in Britain were poor except those of Bates stock, which only wealthy people could afford.[38]

The Scottish farmer faced an even more daunting task than had Bakewell, the Collings, or Bates. Cruickshank had to create a fixed type within a gene pool of animals already bred to provide some consistency. He bred, in effect, to redefine a purebred breed. While he was determined to work within the purebred system by making sure his stock held pedigrees in Coates's Herd Book, he did not wish to use inbreeding, especially to excess, to achieve his ends. He found quite early that he could breed cattle that he saw as desirable without using fashionable stock or inbreeding, but he could not fix that type until 1859, when he produced the bull Champion of England. The animal did not appeal to many cattlemen and did not achieve show success, but it was on him that the Cruickshank herd was built. By breeding Champion of England to his chosen foundation females, Cruickshank found that he could create a short, blocky animal that matured earlier than the tall, refined English Shorthorn of the Bates type, fattened more easily, and bred truly. Ultimately this Cruickshank beef cattle type dominated all British beef cattle breeds wherever they went in the world.

Cruickshank was a master breeder, like the Collings and Bates, and though he dismissed pedigree devotion and the related issue of inbreeding as breeding strategies, his success remained tied to market conditions, as it had for the breeders before him. The chief factor in the ultimate widespread popularity of Cruickshank type cattle was his ability to sell stock in the export trade, and the purebred system with its public registry helped. His cattle did not initially attract the elite and did not do well in the show ring. Thus they did not stimulate speculation as an underpinning to their value until well after they commanded respect in the transatlantic trade with buyers interested in utility farm animals. Cruickshank's large export market began to develop from sales to Canada, and his most important contact would be the Ontario farmer John I. Davidson.[39] Born in Aberdeenshire, Davidson immigrated to Canada in 1842 as a farm laborer. In 1871 he began importing Shorthorns from his native Scotland, and he believed that the Cruickshank type was the way of the future. Davidson determined to import and breed this type of Shorthorn because he thought there would be a market for them in the Midwest, the beef cattle center of the United States. He therefore eagerly imported the new Scottish Shorthorn at the height of the Duchess boom.

Although the Ontario farmer did not actually meet Cruickshank for over twenty years, the two trusted each other completely in business matters and became friends as well. "I cannot sufficiently express my gratitude at what I shall call downright honesty, it is beyond what I ever expected and I hope ever to feel myself grateful to thyself and family for this honorable transaction and I trust we will continue to have similar dealings in the future," Cruickshank wrote to Davidson in 1876.[40] By the early 1880s a demand for Cruickshank cattle had developed, and both the Scottish and Ontario farmers profited from their early conviction that the new type would find popularity with the general farmer. Cruickshank remained loyal to his Canadian friend, and Davidson held almost a monopoly over the importing of Cruickshank-bred stock into North America. In 1888 the Scottish farmer decided to disperse his herd, and he gave Davidson first option on all the animals. "Our connection has been so long standing and so pleasant to me, I could not think of doing a thing so important without consulting thee and giving thee the first offer of what there is to sell. I would much prefer to let thee have the herd as a whole than dispose of it any other way," Cruickshank told Davidson in 1888.[41] The Ontario farmer could not raise the money, and the Cruickshank stock was divided among a number of breeders in Britain. The type remained in fashion with farmers and ultimately attracted hobbyists as well.

By the late nineteenth century Cruickshank stock drew considerable attention in the show ring, and the tide turned in favor of the new type. The Canadian-bred Cruickshank Shorthorn bull Young Abbotsburn made that clear through his victory in 1890 at the Illinois State Fair, the seat of the North American beef cattle world. Both British and Canadian breeders found they had an almost insatiable market for such stock in the Midwest. While Cruickshank did not believe in excessive inbreeding (he even advised his buyers not to use his stock for inbreeding), by the end of the nineteenth century men who acquired his stock within this new market began to practice inbreeding on it. Pedigrees became important as well. Cruickshank stock became known by families, or "tribes," that had names like Lavender, Broadhooks, Orange Blossom, Crocus, Lovely, Spicy, Rose, Butterfly, and Fragrance. Notions of purity became attached to these tribes in a way that the Scottish breeder never had intended. While by the late nineteenth century pedigree and inbreeding expanded as part of the production of valuable Cruickshank stock, one other "craze" in Shorthorns dovetailed with these considerations. A desire for a certain color arose, regardless of quality and even outside pedigree issues. Canadian and British breeders found they had to supply their American buyers with red cattle. White became completely unacceptable; roan was not wanted until the beginning the twentieth century, and even then it never achieved the popularity of red. The American desire for red influenced the market for Shorthorns everywhere. Even Cruickshank, near the end of his breeding career, had to yield to the American desire for red, much as he believed that color could not be related to quality. In 1882 he reported to David-

Young Abbotsburn, a champion Shorthorn bull bred in Canada. The great winner of the 1890 show circuit in the American Midwest, the bull represented a new type of beef Shorthorn that had been developed in Scotland by Amos Cruickshank. Blocky and short, as suggested by his massive head, this bull presents a face full of kindness and an almost human expression in his eye. These features imply tractability, a characteristic important to those handling a bull. *American Agriculturalist*, November 1893, front cover.

son that he had sold a white bull but that the animal "would have brought considerable more if red or roan."[42] In 1887 Amos's nephew informed Davidson that at Sittyton that year there had been a calf crop of fifty-eight, with no whites and only fifteen roans and that there should, as a result, "be a good many which should suit your market."[43]

Canadian breeders complained about American breeders' allegiance to red cattle. The *Canadian Live-Stock and Farm Journal* claimed that a red bull would sell in Chicago regardless of quality.[44] By 1885 Canadians found they could not sell white bulls to anyone.[45] White animals could not be used for breeding, so breeders castrated purebred white calves of excellent quality and showed them as market steers, thus creating a desire for the color in that market. One very significant white Shorthorn steer in the North American show world was Clarence Kirklevington, bred at Bow Park, Ontario. An Ontario breeder-importer of Cruickshank type cattle commented on the red craze in 1902: "Dark red ruined Cattle in the U.S. for many years, in the days of what now is Called the *Color Craze.* Dark red was always unpopular in Scotland." Red Shorthorns "have the worst hair & the thinnest flesh," he added.[46] Another Ontario breeder, who had spectacular wins at the Chicago Fair of 1893, found he could not sell his cattle because of rumors that he had white stock in his herd.[47] In light of this fascination with color, it is interesting to note the color composition of Bates's famous herd when it was dispersed. Most were roan, with a few reds and some whites. While over Bates's breeding lifetime the prevailing color of the Duchesses had been red-and-white, when the herd sold that day, only fifteen were red-and-white, twelve were red, thirty-eight were roan, and five were pure white.[48]

It would take the new science of genetics to show that red and white merely reflected the two extremes of roan. Roaning results from an inter-mixing of white and red hairs. White animals merely have few red hairs, and red animals have few white hairs.[49] Shorthorns had naturally red, roan, and white color patterns—just as Bates's herd demonstrated at the time of his death. Faddism for color did not vanish with the understanding of roaning. By the mid-twentieth century, some hobbyists sought out white specifically and kept herds composed entirely of white Shorthorns.

International trade in the Shorthorn world over the nineteenth century made it imperative that formal structures be put in place to govern record keeping. The regulation and control of pedigree standards for public record books developed coincidentally with the growth of trade between Britain and the two North American countries, as many scholars have noted.[50] Public record keeping proved to be a complicated process in it-self, though, and could reflect a variety of standards. While North Amer-ican breeders were the first to be concerned with the problem, their British counterparts in the end found that they too had to define what the stan-dards should be for entry in a registry (generally the length of the pedi-gree or the number of generations recorded). Coates's book did not ini-tially attempt to set the number of generations needed for pedigree status, because the genealogical background of many animals was unknown. Also, breeders generally were unconcerned with pedigree length.

British breeders did not hurry to change that situation by setting standards for entry. Because Shorthorns were both indigenous to Britain and numerous there, rigid standards made less sense than they did in North America, where the cattle could be described as exotic transplants and were relatively rare. Pedigrees in Britain did not have to "certify" that an animal was a Shorthorn in the same way. As a result, in the 1830s most British Shorthorn breeders did not register their cattle in the public herd book. They often preferred to maintain their own pedigree records or else kept no records of breeding at all. Length of pedigree, even in the earliest of Coates's books, reflected this multiplicity (private and public) of pedigree keeping.[51] Some pedigrees in Coates's book might be as short as one generation, but they could be as long as seven or eight—depending on whether the breeder registering the stock had kept lengthy personal records. By the 1850s the idea of quality had become linked to the notion of purity, but many breeders, especially those not involved in the transatlantic trade, did not use Coates's Herd Book. As late as 1887, numerous Shorthorns in Britain did not have public, or registered, pedigrees.[52] But as the century wore on, the connection of pedigree to both purity and success in the transatlantic trade made Shorthorns with public pedigrees more valuable than those without and encouraged the spread of public registration.

Although Shorthorns existed in North America by the beginning of the nineteenth century, the import movement did not gather critical momentum until the 1850s. Increased importing emphasized the need for a formal structure for pedigree keeping and for pedigree standards. That Shorthorns were less numerous in North America than in Britain made Americans and Canadians view pedigree standards differently than did British breeders. The North American countries tended to look on importing in one of two ways. Some importers argued that incoming Shorthorns should be seen as cattle that could be grafted on to local animals or be used over generations to "breed up" to improve local stock. Others believed the new cattle should be kept separate from indigenous animals—that the blood of the pedigreed creatures would be contaminated by that of the local herds. It became a question of how much, if any, breeding up (over how many generations) should be required to acquire purebred status.

Two viable systems of pedigree qualification developed and competed for recognition according to these two outlooks. In effect, each gave a different definition of what animals could qualify as "purebred." One, known as the four-cross system, made animals eligible for pedigree registration if they had been "bred up" from local stock by using imported, registered purebred males over four generations on the progeny that resulted from the crosses. The other, known as the closed herd book system, granted pedigrees only to animals descended on all sides from imported, registered stock. Early in the nineteenth century the four-cross system generally was accepted as the sensible standard, but its desirability declined over the years in both countries, probably for two reasons. First, Bates had convinced people that purity could be reflected in pedigrees, suggesting that an animal's quality could be defined by the standards of its pedigree. Second, as Shorthorns became more numerous, a sizable market could be sustained for purebreds that had descended from imports alone.

Regardless of standards, public recording required a structure for compiling pedigrees into books and a body to regulate that structure. From the beginning of public pedigree recording in North America, the United States and Canada followed different routes. In the United States, the British pattern developed. Public pedigree recording was undertaken by individuals or private groups (as opposed to government), who set the standards for entrance to the book and ran the system for their own benefit. Breeders had no body or association to speak for their overall interests until later in the nineteenth century. In 1846 Lewis F. Allen of New York State released a small herd book for Shorthorns. As in Britain with Coates's book, Allen's book was a business venture. The pedigrees he collected tended to be for those animals he knew firsthand and pertained only to Shorthorns in New York, Pennsylvania, and New England. By 1855 Shorthorn breeders in the Midwest also registered in Allen's book, named the American Short-Horn Herd Book.[53] Dissatisfaction with Allen's book owing to spurious pedigrees soon triggered two other public Shorthorn herd books: one in Kentucky, created in 1869 by A. J. Alexander and known as the American Short-Horn Record, and one in Ohio, founded for local use by some Ohio breeders in 1878 and named the Ohio Short-Horn Record. All three books developed as privately owned ventures that set requirements for entry and ran the registries for profit. For example, by the 1870s Lewis Allen courted the Canadians for business by saying that breeders in that country had a ready market in the United States and that registry in the American book made it easier for Canadians to sell cattle in the United States.[54]

In Canada a different type of system took root, similar to what subsequently evolved in Europe. From the beginning, government ran public pedigree keeping and maintained the herd book. The first Shorthorn pedigrees existed within central Canada by 1854 under the auspices of the Agriculture and Arts Association, a voluntary organization (founded in 1846 and funded by the government) that ran all agricultural activities within the part of the United Canadas colony known as Canada West (later Ontario). These pedigrees were not numerous enough to warrant publishing a herd book until 1867.[55] The Agriculture and Arts Association, not the breeders as a group or a private individual, also set the requirements for pedigree status. Unlike the American or British system, then, in Canada the government controlled the herd book for Shorthorns and funded it as well. Canadian breeders at that time had no body to look after their collective interests.

By the 1870s both British and North American breeders of Shorthorns began to take a more active part in running registries. A prerequisite to any shift toward breeders' control of recording, however, was for them to amalgamate within associations. Shorthorn breeders in Britain were the first to do so, and at a convention held in 1874 they established an organization called the Shorthorn Society of the United Kingdom and Ireland. The society then set out to acquire Coates's Herd Book. When that registry was founded in 1822, various individuals had advanced funds to Coates, under the agreement that all incoming revenue generated by the registry was to belong to Coates or his successor once proceeds from book subscriptions had covered the loan. (This does not yet seem to have happened as late as 1836.)[56] Coates (and later his son) published books 1–5 between 1822 and 1846. In that year the Coates family sold the registry to Henry Strafford, who put out volumes until 1872.[57] In 1874 the new society bought Coates's Herd Book from Strafford.

After the purchase, the society controlled how the book would be published, regulated costs and profits, and could set standards for entrance. For the moment, standards remained what they had been. Cattle could enter Coates's Herd Book with one- or two-generation pedigrees and often with only recording in the male line. By the end of the century, North American breeders put increasing pressure on the society to change these standards, because no guarantee of purity existed with unknown female lines. British breeders ultimately complied for market reasons, not reasons of quality.

The situation for Shorthorn breeders proved to be more complicated in the United States than in Britain because there were several public herd books. In 1872 American Shorthorn breeders held their first meeting, called to deal with herd books, spurious pedigrees, and standards. The breeders managed to persuade Lewis Allen to adopt and also enforce the closed herd book system. Multiple books, however, undermined the control that American breeders had over all registry issues, and in 1876 breeders formed what would be called the American Shorthorn Association to solve that problem. They set out to take over pedigree management by buying the books and amalgamating them into one. The association purchased the Ohio and Kentucky books for $13,000, and in 1883 it bought the American Short-Horn Herd Book for $25,000. From that point on, breeders had full control of pedigree matters and could regulate expenses and profits.[58] To prevent entry of spurious new pedigrees in the American book, breeders maintained standards on the closed herd book system. This position on standards and the American breeders' greater ability to set them did not end spurious pedigrees or protect the public from fraud, but the new American purebred breeding structure would influence how registration worked in Canada.

Canadian breeders faced different registry problems than did their American counterparts. Because the government of Ontario funded the Canada Short-Horn Herd Book, it was owned not by an individual, a company, or a consortium of breeders but by the government, which also set standards. (Many members of the Agriculture and Arts Association were Shorthorn breeders, but breeders' control of Shorthorn affairs in the 1860s was anything but direct.) Canadian breeders imported from Britain, but they also exported significant numbers of cattle to the United States. Expanding beef operations in the American Midwest drew extensively on purebred Canadian cattle genetics from the 1860s into the twentieth century. For Canadians, standards had to make sense with respect not only to importing cattle from Britain but also to exporting them to the United States. The export market became so important to some Canadian breeders that it influenced their breeding decisions. The production of red Shorthorns illustrates this trend. Farm journals in Canada deplored the influence of market demand for red on the pursuit of real quality in cattle breeding. Reporting on sales in 1909, the *Farmer's Advocate* noted foreign buyers' preference for red, and Canadians' desire to supply it. This situation demonstrated the strength of the almighty dollar, the *Advocate* warned, because everyone knew that red was the worst color in Shorthorns.[59]

Many Canadian Shorthorn breeders found that their main markets were in the United States, not in Canada.[60] They often enhanced their export potential by registering their animals in one of the American herd books, as shown by breeders' testimony to the Ontario Agricultural Commission in 1880.[61] Entry in a Canadian book could also encourage export if the standards matched those in the United States, and that fact made breeders organize to take control of registry issues from the Agriculture and Arts Association.

The Agriculture and Arts Association had adopted the four-cross system in 1867 when it published the first Canada Short-Horn Herd Book. More Shorthorns and better Shorthorns in Canada made the government body consider changing the standards by increasing the number of generations needed to breed up local stock to purebred status.[62] Friction developed between breeders and the association when the organization decided to stay with the four-cross system. That action simply shut down United States markets for cattle with Canadian pedigrees. "Our American cousins ceased to recognize our herd book altogether, insomuch that our Shorthorn breeders who were looking to the splendid markets of the west as an opening for their cattle were necessitated to register only in the American Herd Book," noted the *Canadian Live-Stock and Farm Journal*.[63] Even if Canadian breeders did register in American books, the general low opinion in the United States of the Canada Short-Horn Herd Book hurt Canadians' national pride. Breeders did not like to see the Canadian pedigree system perceived as inferior to that in the United States.

Important breeders denounced the way the Agriculture and Arts Association ran the herd book.[64] Farm journals called for control of both the book and standards by the breeders themselves.[65] In 1881 prominent Canadian Shorthorn breeders met to discuss the issue, and in 1882 they formed the British American Shorthorn Association, which began publishing the British American Shorthorn Herd Book on the closed herd book system. These breeders hoped that the Agriculture and Arts Association would hand over responsibility for the Canada Short-Horn Herd Book so that the two registries could be combined. No one wanted two systems.[66] The Agriculture and Arts Association refused to do so, claiming that it supplied a service to many Canadian Shorthorn breeders. A battle raged between the two groups until 1886. By that time the position of the Agriculture and Arts Association no longer commanded any respect among breeders, and the British American Shorthorn Association was so short of funds that it could not function.

In 1886 the two groups agreed to establish one herd book called the Dominion Shorthorn Herd Book, under the closed herd book system but published and managed by the Agriculture and Arts Association. The breeders had set standards to make cattle under them eligible for entry into the American Short-Horn Herd Book.[67] Breeders also formed a new association called the Dominion Shorthorn Association. "The 12th of

January was a red letter day for the Shorthorn breeders," the *Canadian Live-Stock and Farm Journal* stated with relief in February 1887, "and one that we venture to predict will exercise an important influence for good upon the Shorthorn interest of the whole Dominion, for on that day the details of union between the two hitherto rival books were decided upon."[68] Thus the structure of registry control that developed in Canada was different from that in either Britain or the United States. Breeders set standards, but the government published and certified the pedigrees.

Amalgamation of the Shorthorn books, however, caused difficulties for some Canadian breeders. Distressing information arose out of herd book discussions. Some of the best Shorthorns in the country traced to an excellent bull named Roger, who *might* have descended from the 1817 imports into Kentucky. The notion of "impurity" in the Seventeens (because of their lack of public pedigrees) revisited these Canadian breeders seventy-five years later. "Thousands of the best animals in the land will be ruled out on account of this one doubtful pedigree alone, and some magnificent herds will be almost entirely obliterated," mourned the *Farmer's Advocate.*[69] Reluctantly, breeders decided that Shorthorns tracing to Roger should not qualify for the Dominion Shorthorn Herd Book, because all animals eligible for pedigrees had to trace to imported stock registered in Coates's Herd Book. That decision wiped out some farmers. Having no pedigree status, their cattle lost a large part of their monetary value when they could not be defined as "purebred." Agricultural operations completely changed on some farms over the late 1880s and 1890s, when breeders could no longer designate their good Shorthorn herds as purebred.[70] Under these conditions pedigree status proved fundamental to monetary value and could dictate how farmers operated.

In Britain, Canada, and the United States, purebred animals traditionally could be imported duty free for breeding. But the question would become—especially in the two North American countries—how one defined a purebred animal. Pedigrees became crucial to this issue, because ultimately standards for pedigrees dictated tariff status. Although Canadians made their standards match those of the American breeders, that action did not guarantee duty-free entry because of circumstances that related both to American breed associations and to tariff regulations.

By the late 1880s, Shorthorn breeders in Canada felt assured that their interests had been preserved when a new registry system was established under standards they controlled and set to match those of the United States. Breeders in the United States, however, believed that Canadians undermined American domestic markets for purebred stock. The American

Shorthorn Association asked the United States Department of Agriculture to make registration in the American book a prerequisite for duty-free entry. In 1892 the secretary of agriculture explained the situation to Canadian breeders. "[On] the subject of recognizing the Canadian Record Books in the regulations for the importation of animals duty free, I would say that these books were intentionally omitted. A great majority of breeding associations of the United States have expressed a desire that no books of record should be recognized except those maintained by our own associations." The secretary added that this stance did not reflect on the quality of standards of the Canadian books. It simply represented the unanimous wish of American breeders to enforce registration in the American books.[71] Spurious pedigrees still existed in these books, however, and Canadians claimed to have higher standards than those in the United States.[72]

The Dingley tariff of 1897 reinforced the position of American breeders on importation requirements. American tariff regulations had become increasingly protectionist for commercial (but not purebred) cattle in the late 1880s, and the Dingley tariff further restricted the movement of commercial stock into the United States and set up a structure to control the flow of purebred cattle. The Treasury Department, under the authority of the new tariff, required the United States Department of Agriculture (USDA) to define what animals qualified as duty-free. The Bureau of Animal Industry (BAI), the branch of the USDA responsible for the livestock industries, at first allowed duty-free entry if the animals carried a pedigree from their place of origin. Shorthorn breeders restricted entry by initiating their own form of protectionism. In 1901 the American Shorthorn Association demanded that imported animals have American pedigrees in order to be recognized by American breeders as purebred for breeding purposes, and then they charged $100 to register non-American animals. Canadian and British breeders saw the fee as a tariff or tax.[73] Breeders who hoped to export to the United States also viewed the action as merely a moneymaking effort by the American Shorthorn Association.[74] Public herd books could apparently be used as agents to control markets even when pedigree standards remained identical. Many Canadians argued that American Shorthorn cattle should not be able to enter Canada so freely. "The ease with which any kind of animal with a pedigree can be brought into Canada from south of the line, and the almost prohibitive character of the regulations with which the United States government controls similar trade into their country, is not at all credible to our self respect and independence," noted the *Farming World and Canadian*

*Farm and Home.*[75] Breeders in Britain also were incensed and saw the move as a general trend toward keeping out imported cattle and at the same time generating money for the American Shorthorn Association. "Protectionist Shorthorn intriguers must needs [be] more Protectionist than their own Government, and hence the imposition of the stupid registration fee in question." Clearly, this journal noted, the action protected the value of American stock.[76]

In 1906 the protectionist actions of the American association received support from the USDA. The Bureau of Animal Industry, under BAI order 136, which replaced order 130, altered importation rules. After 1 July 1906, all imported purebred animals had to be registered in the American books to obtain duty-free status.[77] This action gave legislative force to the regulations of the American Shorthorn Association. The news angered Canadian breeders even more.[78] The Canadians retaliated by asking the Dominion government to restrict the duty-free entry of Shorthorns from the United States not registered in the Canadian book.[79] But the whole issue had become more complicated than a matter of North American tariffs.

The American-Canadian Shorthorn herd book problem and importation issues did not result simply from conflict over markets. First the Americans and subsequently the Canadians became increasingly unhappy with pedigree standards set in Coates's Herd Book. American breeders also noted that the pedigree status of many Canadian cattle reflected the British standards in that registry, where pedigrees continued to be short. The British Shorthorn society refused to see that this situation presented a dichotomy: stock that could obtain pedigrees in Britain would not have qualified in North America even under the four-cross system. Shorthorn breeders in Britain became more receptive to demands for standard changes, however, when they recognized a threat to their markets. That situation began to develop as early as the 1880s. After the North American countries had formally moved to the closed herd book system, the British system made little sense in light of the lax entrance requirement of the stock to be imported. From the North American point of view, British standards did not reflect what pedigrees should guarantee—purity and, through purity, quality. In 1900 American breeders asked the government not to accept registration in Coates's Herd Book for duty-free status.[80]

The problems of importation and purity made American breeders ask the Dominion Shorthorn Association in 1900 to act in conjunction with them to press the "Short-Horn Society of Great Britain and Ireland to make a more complex index to their Herd Book and to discontinue recording animals whose dams were not on record." Breeders wanted Coates's Herd Book closed to cattle with any unrecorded ancestry. Richard Gibson, who had been a longtime member of the British society, warned Canadians that the British breeders would not change. Many good Shorthorn cows in Britain did not have pedigrees, he pointed out, and females could not be registered in Britain until they had calved. The Dominion Shorthorn Association, however, agreed to support the American endeavor and to send delegates to Britain to discuss the issue along with those from the United States. Canadian breeders hoped to induce their British counterparts "to record in the future all their dams, and compile a more sensible index to their Herd Book, and to discontinue recording Short-Horns whose dams [were] not recorded."[81] The meetings went well, and all three national associations decided to streamline their standards as soon as they could work out how to do pedigree listings in a more unified way.[82] Duty-free entry did not follow immediately, as we have seen, but at least breeders agreed that standards in each country should be acceptable to breeders in the others. An active move to pedigree reciprocity developed in 1911, when the import regulations of the Bureau of Animal Industry were revised. Henceforth purebred animals, if certified by the BAI, no longer had to be registered in the American books. They needed pedigrees only in their country of origin.[83]

The American Underwood tariff bill of 1913 enhanced the move to pedigree reciprocity. It allowed all cattle into the United States duty free. Shorthorns, then, no longer needed purebred status to avoid duty.[84] This situation weakened the American breeders' ability to impose restrictions on Canadian Shorthorns registered under standards identical to their own. Although the United States removed its tariff in 1913, Canada continued to monitor entry of purebred stock, but it did recognize foreign herd book standards for duty purposes. Animals could be imported duty free only by people living in Canada.

Government supervision of the purebred industry in the United States had existed for a short time before the new regulations of the Underwood tariff took effect. The tariff of 1897 had authorized the Bureau of Animal Industry to regulate the American purebred animal industry, and in 1904 the bureau took greater control of American pedigree-recording associations. The new regulations for American breed associations were laid out in BAI order 130. "These regulations require," the bureau reported in 1905, "for each certified American Association an annual report, the publication of books of record, and their submission to the department for publication. The associations must submit complete sets of the published volumes of their books of record, and complete statements of their business methods and financial condition."[85] In 1908 the bureau claimed that the purebred breed associations in the United States liked this supervision of their affairs and that the greatest benefit of the tariff regulations was the internal control of the purebred industry. "It is rather anomalous," the bureau noted in that year, "that a law intended to guide customs officers in deciding what animals should be imported duty free for breeding purposes should in its operation be so changed in effect that it is of much more importance to the country as a measure for the supervision of local record associations."[86] The bureau made it clear that only a tax act protected breeders from fraudulent American pedigree keeping. The USDA suggested that Congress empower the department to control purebred associations directly. This did not happen, and the 1913 Underwood bill, which removed the tariff, effectively undermined the BAI's authority over the American registration of the purebred breeds.

Meanwhile, in Canada there were new developments in herd book control and the running of breed associations. In 1895 the Agriculture and Arts Association was abolished. The Shorthorn association subsequently had complete control of its herd book—a situation identical to that in America. Dominion legislation in 1900, however, initiated a change back to government regulation and control by passing an act covering the incorporation of livestock associations. To incorporate under this act, a breed association had to provide for "the registration of pedigrees of pure-bred stock," set up a constitution with by-laws, and provide the Dominion minister of agriculture with an annual report of finances and business undertaken. No changes in the constitution could be made without the approval of the minister of agriculture. Only one association for each distinct breed in Canada could incorporate.[87] Shorthorn breeders questioned the new act at first, because they feared they would lose control of registry standards, an issue they had won in the 1880s with the resolution of the conflict between the Agriculture and Arts Association and the British American Shorthorn Association. But by 1901, after breeders' fears had been put to rest, the breed association had incorporated under the act.[88]

Conflict over railway rates for moving purebred cattle to western Canada and the impasse with respect to duty-free importing of Canadian cattle into the United States sparked greater government involvement in Shorthorn registry. In 1904 railways refused to carry purebred animals at reduced fares unless they had a pedigree guaranteed by the Dominion Department of Agriculture to be correct. At the same time, many breeders believed that "it [was] time some action was taken by the Dominion Government in regard to regulating the importation of pure-bred stock into Canada."[89] Breeders called a national livestock convention in 1904 and passed a resolution "favoring the nationalization of all records for pure bred stock kept in Canada, and asking the Department to take the necessary steps to bring it about."[90] In 1905 the Canadian Livestock Records Corporation, under the supervision of the Dominion Department of Agriculture, began acting as a national registry for Canada's purebred stock.

In 1912 purebred livestock affairs in Canada received more specific regulation by the Live Stock Pedigree Act, which amended and replaced the earlier act of 1900. "The registration of pedigrees is left within the powers of the association, the ruling being set forth clearly what animals are eligible for registration," the *Farmer's Advocate* explained.[91] The formation of the Canadian Livestock Records Corporation in 1905 and the Live Stock Pedigree Act of 1912 set up a system for regulating the purebred industry in Canada that was unique in relation to what had evolved in both the United States and Britain. Government played an active role in governing the breed associations and certified their records, but breeders retained complete autonomy with respect to standards.

American breeders noted the creation of the Canadian Livestock Records Corporation with interest.[92] Many envied the supervision that the Canadian government provided the purebred industry through the Dominion act for incorporation of livestock associations and the national registry system. An American farm journal, the *Wallace Farmer*, commented: "It would have been worth an immense sum to the breeders and farmers of the United States if the matter of recording pedigreed stock had been placed in the hands of the government twenty-five years ago. We do not know of anything that would do more to benefit the pure bred stock breeding at the present time than for the United States to enact just such a law as is proposed in Canada." The journal regretted that in the United States many breed records could and did exist for each breed, that no specific legislation for supervising them had been initiated, and that as a result many spurious records could be found. Any person could start a registry at any time under any conditions.[93] No wonder the Bureau of Animal Industry had been delighted with the action of the Treasury Department. Indirect legislation gave the USDA at least some supervision of herd books and thus some control, even if neither lasted.

It is worth speculating on why the Canadian system developed differently from its American and British counterparts. One reason might be the underlying "red/tory" political culture of Ontario. Government activities there over the late nineteenth century tended to reflect a sense of deep conservatism alongside a leaning to socialism. This mentality presented a strange right/left attitude toward the role of government. A deeper and more practical reason for government interference in the purebred industry seemed to be the small size of the Canadian industry in relation to its competitors and market partners. There is evidence that the Canadian purebred sector was not large enough to generate the funds needed to run a viable regulatory structure. The declining fortunes of the British American Shorthorn Association and its herd book, for example, related directly to finances; members claimed that the venture collapsed for lack of money.[94] Subscriptions to the herd book were probably too small to support its publication. Funding by the government brought with it a form of regulation that also helped win respect on the international market.

Government control of purebred matters also concerned people in Britain at this time. In 1912 a British parliamentary commission called together the producers of various livestock breeds to discuss the nation's export trade in purebred animals. Fundamentally, the breed associations in Britain acted independently of government control, and one major question brought up at the meetings was whether parliamentary involvement in the regulation of the purebred industry would help international trade. Maintaining public pedigree integrity became central to the question of government supervision. Overall, breeders seemed to think the various societies had maintained credibility by policing their own ranks and by privately punishing culprits who publicly perpetuated spurious pedigrees.[95] No legislation granting specific government control over the British purebred industry resulted from the commission's work. The purebred situation in Britain resembled that in the United States after 1913.

In 1914 world war broke out, and even though the United States did not enter the conflict until 1917, the movement of purebred livestock across the Atlantic to either North American country basically shut down. The war affected how the international purebred animal trade functioned after peace in 1918. General depletion of European livestock in the war, and greater production in North America, changed the way the entire industry worked. The decrease in Europe was less than expected, and the North American countries increased production out of proportion to Europe's needs. In France cattle levels dropped 10 percent, in Germany, some 17 percent (the greatest drop). In Britain cattle stock increased 1 percent. Cattle numbers rose 20 percent in the United States and 54 percent in

Canada.[96] With decreases in Europe and high prices, farmers culled their poorer stock and moved to purebred animals. The same story was true in France, Belgium, Switzerland, the Netherlands, and Britain. Demand for purebred cattle stimulated European production and provided breeders there with a market to replace the old North American one. A further incentive for the European breeders came from the enormous demand for purebred breeding animals in South America, particularly Brazil and Argentina.[97] All these factors helped bring about complete reciprocity of purebred status between the North American countries and Britain. In 1918, for example, the Dominion Shorthorn Association managed to arrange registration of Canadian-bred animals in the American books for the customary fee of $1.25, and American-bred cattle could be registered in the Canadian book for $1.00.[98]

Shorthorn history over the nineteenth and early twentieth centuries reflected British and American concern for better beefing cattle. British breeders succeeded in creating a breed of cattle first wanted at home and then greatly desired in the United States. American demand helped fuel Canadian interest. Breeding strategies were molded by that transatlantic connection, which in turn shaped both pedigrees themselves and perceptions about them. The introduction of public pedigree to inbreeding theory received the most attention from Thomas Bates. But once he had linked the idea of pedigree and inbreeding to quality, men who acquired his stock after his death carried that linkage to an extreme that the master breeder had not intended. The commodity became increasingly scarce, triggering speculation. The inevitable crash of Duchess cattle did not completely remove the linkage of pedigree and inbreeding to quality, particularly in the minds of the elite hobbyists who had been drawn into the cattle market as the boom grew. The boom, however, stimulated a corrective factor even before it reached its peak. Farmers tried to recover the cattle for practical uses. These newer breeders, under the leadership of the Scottish farmer Amos Cruickshank, cared more about fixing type than maintaining purity. Once the new style had conquered, however, something of the old sense of purity and of the importance of pedigree and inbreeding reemerged. The return of the hobbyists to that market introduced faddism again, but never to the same degree as under the Duchess phase.

The transatlantic trade in purebred Shorthorns stimulated the rise of well-defined standards for public pedigrees and of breeder organizations to govern those standards. The market played a critical role in the way the system worked and in the development of its culture. The drive to clearly define pedigree standards evolved in the importing countries of North America. The attachment of purity to pedigrees became more of an obsession in North America because pedigrees were expected to reflect quality in the stock to be imported. The transatlantic trade in Shorthorns made it increasingly difficult for contemporary breeders to separate two mechanisms—breeding tool and marketing device. The connection between pedigrees as marketing agents and pedigrees as breeding tools became almost seamless. Because the division between the two became blurred in breeders' mind, the story of pedigree regulation for Shorthorns over the nineteenth century shows how people related standards, purity, and quality to market conditions. The story also shows that problems arose over how to regulate both standards and markets, and over who should regulate them. While both American and Canadian Shorthorn breeders touted standards as a barometer for purity and quality, in reality standards reflected the patterns in the international market. The United States was the chief driver of the standards issue and the chief buyer of the British product. The way standards came to be defined there influenced how they developed in both Britain and Canada. British standards shifted to match the demands of the international market in North America. Canadian standards reflected American ones as closely as possible.

Organization and regulation of pedigree differed between the United States and Canada. The American system resembled the British system; breed organizations were not subject to direct government control. Historically the Canadian purebred industry was regulated in some fashion by government. By the early twentieth century a complicated pedigree system had evolved in Canada. The national government certified the registry and the government supervised breed associations, but breeders maintained full control of pedigree standards. Legislation in Canada also supervised the way breeder associations functioned. Canada's position within the international market reflects the difficulties that countries outside Britain and the United States had when they played a significant role in the early purebred breeding trade.

In spite of the difficulties that markets and pedigree issues created for the breed, Shorthorns dominated the purebred beef cattle industry in both Britain and North America until well into the twentieth century. The importance of Shorthorns to the breeding of purebred cattle cannot be overestimated. The Shorthorn would become the foundation of over thirty breeds found all over the world. Shorthorn type also influenced the style of other breeds. Short, stocky cattle of the Young Abbotsburn sort came, for example, to dominate the ranks of Herefords and Angus cattle on both sides of the Atlantic. The fortunes of the Shorthorn went into decline in Britain and North America when a revolution in beef cattle breeding evolved in the 1960s. Short, blocky cattle became less popular as breeders sought taller animals with less fat marbling. The shifting preferences coincided with the introduction of artificial insemination, a technology that eased the spread in the late 1960s of European breeds such as Charolais, Simmental, and Limousin. Within the shifting cattle environment, Hereford and Angus cattle managed to hold their own as breeders began selecting for taller and leaner types. The Shorthorn, however, did not recover its favored position. In the 1980s some breeders in North America attempted to regenerate the breed by introducing the genetics of Maine Anjou, a large French breed derived from a Shorthorn base. But the beef Shorthorn did not regain its hegemony.

# Three

## PRODUCING BEAUTIFUL DOGS

"The owners of great herds of shorthorns have spent fortunes to bring out their ideas of what perfection should be, and it may be said that this sort of fascination to excel in the culture of animals pervades the whole country. It is not remarkable, therefore, that the animal which is *par excellence* the friend of man should share with others in becoming the object of serious attention from those who would be inclined to work out improvement on the lines of science." So wrote Sewallis E. Shirley, first president of the British Kennel Club in 1880.[1] He meant the dog, of course, and he hoped to encourage the development of a new industry by applying purebred breeding to the species. This chapter and the next discuss dog breeding in Britain and North America from the mid-nineteenth century until the present. The evolution of the dog fancy (conformation breeding for show), with its support structures designed to encourage the growth of markets, is the subject of this chapter. The fortunes of one breed, the Collie, help elucidate how certain patterns of the dog fancy developed. Breeding practices and particular dogs within the Collie breed, as well as trade issues between Britain and the United States, are the subjects of chapter 4.

The dog fancy, with its show structure for beauty points (field or herding trials played a less important role in the system), led to the breeding of purebred dogs for sport. But another major and separate dog enterprise also developed over the nineteenth century: the old aristocratic sport of Greyhound racing was commercialized. The two developments need to be distinguished. When the industry of Greyhound racing emerged in the nineteenth century, the structures and breeding aims it stimulated remained different from those that developed to support the dog fancy. Because Greyhounds' breeding reflected racing needs, their production would always be based on utility (in this case speed) rather than looks.

In Britain, regulation of Greyhound racing began as early as the reign of Elizabeth I, and in the sixteenth century dogs belonging to royalty and the nobility competed in formal coursing races for the pleasure of the elite. Greyhound racing and a system to support it as a commercial venture (as entertainment to be watched by the general public) arose quickly after the late eighteenth century. The first Greyhound coursing club opened to the public in 1776, and by the early nineteenth century dog racing had attracted considerable attention. The famous Waterloo Cup began in 1837, and in 1858 the National Coursing Club of England started to regulate the sport. Early races did not resemble those we know today. Spectators often followed the dogs on horseback, and live hares served as lures.[2] The first artificial lure for Greyhound racing appeared as early as 1876, but mechanical lures were not in general use until after 1912, when an American, Owen Patrick Smith, invented a more workable lure that could be used on a track. Registration of Greyhounds in a studbook evolved with the racing industry. Interest in coursing and the linkage of the nobility to Greyhounds meant that some ancestry records had been kept privately as far back as the eighteenth century, but no public registry for Greyhounds in Britain existed until as late as 1882. In that year the National Coursing Club required registration in the Greyhound Stud Book for dogs entering races.[3]

After racing became popular in the United States, a structure for Greyhound racing and registry developed there on similar lines. Greyhounds could be found in the United States by the mid-1800s, but they were not used for racing. They were part of breeding programs designed to produce dogs that could hunt wildlife destructive to agriculture, especially in the West. Racing became established in the United States in the twentieth century. The National Greyhound Association, founded in 1906, still runs the registry for all racing Greyhounds in North America. Greyhounds have continued to be bred primarily for racing in both the United States and Britain. Even though the breed is accepted for conformation exhibition, the racing studbook provides pedigrees for the greatest number of Greyhounds in either country. Only with the advent in the late 1980s of organizations that place retired racing dogs in homes have Greyhounds played any substantial part in the dog world outside racing.[4]

A general concern with beautiful dogs arose in Britain in the nineteenth century, at the same time that the Greyhound industry got started. It must have at least partially reflected Queen Victoria's passion for the animals. All her life she loved and collected dogs, taking a personal interest in their breeding. Many people found that the best gift they could present to the queen, who did not lack worldly goods, was a pair of puppies. By 1841 she owned so many dogs that she had a kennel built at Windsor. The magnif-

Portrait of a Greyhound. Eos belonged to Prince Albert, husband of Queen Victoria. By the time the dogs came to serve a dog racing industry in the nineteenth century, royalty had treasured Greyhounds for hundreds of years. *Eos*, by Edwin Landseer, 1841, oil on canvas, 44 by 56 inches. Collection of Her Majesty the Queen.

icent kennels featured clean yards, sleeping quarters with steam heat, automatic water bowls, and freshly cooked dog food. Her interest increased when the sport of breeding purebred dogs and showing them for beauty got under way. Queen Victoria kept her own records for the dogs, and she designed rooms to hold trophies and dog portraits as well as space where she could view any animal she wished. Not only did she supervise the actual breeding of her dogs, she named every puppy. While her favorite breed was said to be the Collie, she also loved Skye Terriers, Scottish Terriers, Irish Terriers, Spitz breeds, and Pomeranians. Other breeds, such as Tibetan Mastiffs, Chows, Saint Bernards, Great Pyrenees, Chinese Spaniels, Pekinese, and Greyhounds also lived at the Windsor Kennels.[5] The Prince and Princess of Wales loved, bred, and exhibited purebred dogs as well. Borzois were a particular favorite of the Princess. Royal support of purebred dog breeding and showing did much to make the whole endeavor fashionable with the landed classes, the nobility, and the middle class.

The earliest showing of fancy dogs seems to have been a sideline to canine rat-killing competitions. By the mid-nineteenth century, owners brought their pets together to be admired in informal exhibitions on nights when regular rat-killing competitions were not held. In 1851 The *Illustrated London News* reported as follows on a "Fancy Dog Show" at a certain club where sons of the best families were wont to go: "At this place a club is held, by one of the rules which, each member is expected (in fact, we believe, compelled), when he attends, to bring a dog for show, or sale, as he thinks proper; thus ensuring a good show on the club night . . . ; and here may be seen the most beautiful specimens of spaniel, Italian greyhound, and of late years, of the Isle of Skye terrier." "The show dogs, or Fancy Pets, as they are termed, are solely valued for beauty of their respective sort."[6] The dogs, the paper reported, had no utility use. King Charles Spaniels, for example, could not be described as good hunting dogs, yet those of the right size and color could command as much as £200. "Italian greyhounds are chiefly valued for perfection of symmetry and color," not for either intelligence or running ability, the paper added.[7]

Showing dogs for beauty, however, soon became entwined with exhibiting fancy poultry. Attitudes toward breeding fancy poultry would ultimately be similar to attitudes toward breeding beautiful dogs. The historical background to fancy poultry breeding and to the problems encountered in the sport provide some context, then, for critical features that could be seen in the early dog fancy. Controversies that evolved within the early fancy poultry industry also reflected patterns and conflicts that still endure in the world of purebred dogs.

With the end of cockfighting competitions in Britain, a pastime outlawed in 1849, selective breeding of poultry for fighting ability became less widespread. The popularity of other types of poultry breeding, however, gathered momentum in the mid-nineteenth century. Poultry imported from the Far East had introduced new genetic material suited to breeding for beauty, and these strains stimulated the breeding of fancy poultry, giving the sport more breeding scope. The first large formal show for beautiful and exotic poultry took place in London in 1844 at the Zoological Gardens. Queen Victoria became interested in poultry at this time and had a fowl house built at Windsor. She showed Cochins in 1846. Royal support of the fancy increased its popularity with the middle classes,[8] and breeding poultry for beauty competitions began to escalate in England's industrial cities.

At this same time, there was a boom in poultry breeding in the United States. Exotic breeds from the East had been imported to America when they entered Britain, and American breeders began to develop new sorts, such as the Brahma, from the Eastern types. As in Britain, it was beauty or fancy points that interested breeders, not utility characteristics like meat or egg production. In 1849 the first fancy poultry show in the United States took place in Boston. The "hen fever" had begun. The 1849 show lasted three days, and people bought birds at skyrocketing prices. Birds worth 50 cents a pair sold for $13.[9] George Burnham, a poultry fancier central to the boom, saw some humor in this transaction. He wrote in 1855, "As high as thirteen dollars was paid by one man (who soon afterwards became an inmate of a lunatic asylum) for a single pair of domestic fowls." "By the time this fair closed," Burnham continued, "the pulse of the 'dear people' had come to be rather rapid in its throbs, and the fever was evidently on the increase. Fowls were in demand. Not *good* ones, because nothing was then said by the anxious would-be purchasers about quality."[10] The boom accelerated after these events, and Burnham played a role. In 1852 he packed up a pair of birds and sent them to Britain addressed to the queen. The *Illustrated London News* covered the story, which triggered still more interest in poultry.[11] Burnham received requests from both British and American fanciers for pairs at prices up to $150. At the height of the hen fever the price went as high as $700 a pair. The bubble burst in 1855. "Never in the history of modern 'bubbles,' probably did *any* mania exceed in ridiculousness or ludicrousness, or in the number of its victims surpass this inexplicable humbug, the 'hen fever,'" Burnham wrote about the frenzy.[12] "What a price to pay for a hen and rooster!" exclaimed the *Farmer's Advocate* in Canada in 1869 some time after the fever had been spent.[13] But bursting the bubble did not end the devotion, and large shows continued to support breeding for beauty. In 1878 the *Canadian Poultry Review* explained the magic of the fancy to young potential fanciers. "Do not make the almighty dollar the sole result to be obtained. Look beyond; get at the science, the nature of things. You have a wonderful power given you. You can fashion and form and color a bird or beast almost as you will, but you cannot make one atom of its texture. Given the crude materials of the world, you could not fashion a feather; you might picture one, but as the picture differs from the reality, so your work in all things falls short of nature. You can *take* the breath from life, but you cannot *give* it. You *mold* but you cannot *create*."[14]

Poultry breeding, of course, produced birds for agriculture as well as for beauty. Poultry farmers and poultry fanciers interested in better output by farms tried to define the relation of fancy to utility in the breeding of any poultry. Fanciers repeatedly stated that farmers should use

A portrait of fancy poultry. By the mid-nineteenth century interest in poultry breeding centered on raising birds deemed beautiful, not good meat or egg producers. Fancy poultry commanded high prices, and a system of show competitions developed to support the hobby of breeding for beauty. The birds portrayed in this painting represent only one of the many types that could be seen in both Britain and the United States by 1850. *A Silver Spangled Hamburg Cockerel and Hens,* by William Shayer, ca. 1850, oil on canvas. Spink and Son Ltd. Bridgeman Art Library, London, England.

purebred fowl, but they also recognized that the birds did not pay. Sometimes farmers claimed that fancy poultry did pay, but in those cases the farmer apparently was acting as a fancier himself, not breeding for utility. Public funding of poultry shows in North America aggravated the problem. Farmers asked why taxpayers should be asked to support exhibitions showcasing useless fowl.[15] In 1905 the *Farmer's Advocate* in Canada had this to say about the fancy industry and the public purse: "It would be interesting to know how much of [the government grants] went to substantial encouragement of the Canadian poultry industry, and how much to line the pockets of a few fanciers of non-utility breeds?" "We are familiar with the arguments advanced in justification of these prizes for non-utility breeds, that they are necessary to 'draw attendance,' to 'stimulate an interest in poultry,' etc. Passing them by as unworthy of reply, we submit that the explanation of this squandering of money on fads is found in the desire of a few fanciers to enrich themselves through funds ostensibly devoted to agricultural purposes."[16]

The poultry fancy continued to flourish. Not until the early twentieth century would selective poultry breeding be done on a large scale and in an organized way to improve meat and egg output, becoming a viable industry separate from breeding fancy fowl. In the long run the poultry fancy aided utility breeding by introducing general guidelines for selective breeding strategies and by stimulating research on the general care of fowl, such as housing and feeding. Strains of some fancy stock ultimately went into the makeup of the utility poultry breeds in the twentieth century. The poultry fancy did not end when the breeding of poultry for specialization in meat or eggs became better established. By 1912, particularly in Britain but in North America as well, two distinct poultry industries existed—one for fancy and one for utility. In 1916 an agricultural expert in Canada described the transition thus: "The breeding of poultry is old as an art but new as a science."[17]

The problems of breeding for fancy or utility, the meaning of improvement within that framework, and the effects of the marketplace that emerged in the fancy poultry industry would be deeply embedded in purebred dog breeding from the time that that fancy got under way. By the end of the 1850s exhibitions of beautiful dogs often took place in conjunction with fancy poultry shows. The famous Newcastle-on-Tyne show of 1859 in Britain, for example, was part of a large poultry exhibition.[18] Catalogs for British dog shows, even late in the nineteenth century, were often combined with catalogs for poultry exhibitions. In North America the farm press showed the same affiliation of fancy poultry affairs with dog issues. Information on breeding dogs appeared regularly in the poultry section of Canadian farm journals. In 1872 the *Canadian Poultry Chronicle* called for dog shows in Canada to be run in conjunction with poultry exhibitions. Dog fanciers should unite with the Poultry Association of Ontario to achieve this end.[19] In 1878 the Montreal Poultry, Dog and Pet Society was formed, and the organization had its first dog show at the end of that year. The *Canadian Poultry Review* published extensively on dog affairs throughout the 1870s, and the journal established a kennel section in 1885. Many dog articles found in the Canadian poultry press in the 1880s were reprints of features originating in American poultry journals such as the *Ohio Poultry Journal*.[20] In 1889 the *Canadian Poultry Review* enlarged its kennel section by printing the *Kennel Gazette* as a separate journal.[21] The following year the *Kennel Gazette* became independent.

From poultry journals, trading poultry for dogs seems to have been fairly common in the late nineteenth century.[22] Early dog breeders had often, not surprisingly, started out as poultry breeders. Thomas H. Stretch, one of the most significant of the early British Collie breeders in the 1880s, originally bred a variety of fancy fowl.[23] An American Collie breeder and one of the greatest experts ever on Collie history had also been a poultry breeder. In 1903 a country doctor named O. Prescott Bennett became involved with Collies when he traded a breeding pair of birds for a pair of Collies.[24] The fowl-canine connection did not end abruptly at the beginning of the twentieth century. As late as the 1920s some dog show judges were poultry breeders.[25]

After the 1861 Birmingham dog show, the popularity of exhibitions for beautiful canines accelerated.[26] Early shows, however, soon were plagued by favoritism, corruption of judges, and fraudulent handling of the animals. Throughout the 1860s, plucking, dyeing, and clipping of hair deliberately hid flaws, and entries might be falsified as to age and type. By the early 1870s the corruption associated with dog showing and the collapse of the National Dog Club (founded in 1869) threatened to end the embryonic fancy of exhibiting dogs for conformation. The need to regulate these exhibitions gave rise to the British Kennel Club, founded in 1873 in London.[27] The Kennel Club was the first organization of its type to survive, and it was a model for dog associations that followed in the 1880s in other countries. The French and Italian kennel clubs, both formed in 1882, were modeled on the British one, as was the Dutch kennel club that started in 1890.[28] The two major, enduring North American clubs, also initiated in the 1880s, had similar but not identical structures.

Sewallis E. Shirley became the leader of the select group of twelve British men who formed the first Kennel Club. Born in 1844 and educated at Eton and Oxford, Shirley was a wealthy landowner and a member of Parliament. He loved dogs, particularly gun dogs. While a boy at Eton he had kept his own pack of beagles for hunting, and as early as 1865 he had exhibited fox terriers at dog shows. He used dogs for both field trials and conformation competition, but as time passed he emphasized conformation. He and his fellow founders of the Kennel Club intended to keep the organization as elite as possible. Membership was to be limited to one hundred men drawn from the landed classes, nobility, or royalty. The Prince of Wales was the first patron. Many of that first hundred were members of the aristocracy. In 1880 Shirley established the *Kennel Gazette* as a private venture of his own, but the next year he gave it to the club. He served as chairman of the Kennel Club for twenty-six years.[29] The Ken-

A dog show before the kennel clubs. Even though dog showing did not get under way until about 1860, there were get-togethers for exhibiting dogs before
then. Notice that the dogs being compared in this show are not all of the same
breed. *An Early Canine Meeting*, by R. Marshall, 1855, oil on canvas, 29 by 37 inches.
Collection of the Kennel Club.

nel Club continued to restrict membership by class and gender; women
could not join until as late as 1979, even though a Ladies' Branch had existed since 1899 and after 1900 at least half the exhibitors at shows were
women.[30]

Registry issues and breeding practices were not central in the formation of the Kennel Club. The need to regulate entries for specific classes,
however, quickly brought up the issue of maintaining a public record
book. The situation—a need for formal organization of competitive
events—resembled the one that spurred the establishment of the General
Stud Book for Thoroughbreds in 1791, and dog people at the time saw the
parallel.[31] In 1874 the Kennel Club began to issue a public record book,

owned and controlled by the Kennel Club. The book listed dogs only by name. Many early entries had no pedigrees at all, and the foundation animals often were of unknown parentage. Not everyone interested in showing dogs favored a public record book. The London dog men who had set up the Kennel Club and its public book almost immediately came into conflict with the dog interests at Birmingham. The Birmingham Committee violently opposed registering dogs in public books, but it found as early as 1875 that it had to accept the Kennel Club's studbook.[32]

Shirley addressed the topic of public registries and trade in his *Kennel Gazette*. In the first issue of the journal he pointed out the eminence of British breeders with respect to animal improvement and the value that people in other countries set on British stock. Publicly registered pedigrees, Shirley argued, had greatly increased the value of the Thoroughbred and the Shorthorn. "It is now more plainly seen ... that science must be associated in the breeding of dogs in as great a degree as it has been in the connection with racehorses and shorthorns," he wrote, and by "scientific" production he meant breeding by the public pedigree system.[33] "Whereas the General Stud Book has spread the interest taken in race-horses, and the opinion throughout the world of the Herd Book has done the same thing for shorthorns, we can point to the fact that English dogs have increased prodigiously in importance since the Kennel Club Book was started. Our American friends would not have given three or four hundred guineas a piece for setters, and two hundred and fifty guineas for a pointer, unless there had been a public record of their pedigrees," Shirley wrote in 1881 in the Kennel *Gazette,* "and the same argument holds good down to the present moment, when we hear that a setter ... had been sold for a stud in America, for a price nearly good enough for a short-horn bull, or a collie is also sold for another corresponding sort of price."[34]

By 1883 Shirley began arguing that in order to sell dogs at all, breeders needed to provide public pedigrees. "In America, and on the Continent, this feeling in connection with dogs is even stronger than it is in England, as an American or foreigner stands out for a pedigree above everything else." When English dogs did not have publicly recorded pedigrees, as had been the case earlier, they commanded little money either in the United States or on the Continent.[35] British breeders, Shirley made clear, did not value public pedigrees as much as their North American counterparts, but they should take an interest in registry for marketing reasons. Shirley knew that the dog situation in the United States in particular was critical to the British canine industry.

By the 1870s the sport of purebred dog breeding and showing had become well established in North America. New York's still prestigious Westminster Dog Show, for example, started in 1877. The evolution of dog shows in North America brought about the need for regulation, just as it had in Britain, and kennel clubs came into existence to fulfill that need. The National American Kennel Club, founded in 1876, began to publish a studbook in 1879. Patterns seen in the American purebred cattle world repeated themselves. Either individuals or companies, not governments, owned and controlled early public dog registries. Dr. Nicholas Rowe started the studbook of the National American Kennel Club. (The Forest and Stream Publishing Company issued a second public studbook for dogs in 1881, known as the American Kennel Register.) The National American Kennel Club had from its beginning been more devoted to field trials for dogs than to conformation shows, and when the organization changed its name to the National Field Trial Club, dog breeders interested in breeding for conformation began to think of new directions.

At this time breeders in Canada agreed to work with their counterparts in the United States. By the early 1880s there had already been some internal attempts to organize the Canadian dog situation. A group of Canadian breeders had formed the Dominion of Canada Kennel Club in 1883, but it does not appear to have ever set up a registry. Perhaps that was one reason Canadian breeders became interested in movements in the United States. Of the thirteen dog clubs that founded the American Kennel Club in Philadelphia in 1884, three were Canadian. A Canadian served as the club's first vice president.

From the beginning, the American Kennel Club differed in one significant way from its British counterpart: individuals could not be members. Various dog clubs made up the membership. The inherent elitism deliberately embedded in the British system thus seemed less strident in the American club. Also unlike the British club, the new organization planned from the beginning to run a public studbook. When it could not buy the studbook put out by the Forest and Stream Publishing Company, in 1887 Dr. Rowe presented his studbooks to the club as a gift. Shortly after, the American Kennel Club moved its headquarters to New York (where it has remained ever since), and in 1909 it became incorporated in its present form by the state of New York. In American tradition with respect to purebred matters, the club maintained full control of its registry, without direct supervision by government. Rival kennel clubs with registries were started in the United States over the years, but the American Kennel Club managed to maintain its dominance.[36] It could not, however, shut down the multiplicity of record keeping for dog breeds. The United Kennel

Club, for example, a privately owned all-breed registry for dogs founded in 1898 (then bought in 1973 and still owned by an aerospace executive), still functions. It remains the second oldest and second largest dog registry in the United States.

Friction developed between the Canadian and American members of the new American Kennel Club, and in 1886 the Canadians withdrew. That Canadians' interest in dogs had become considerable was evident from the size of exhibitions in Canada. The Toronto Dog Show of 1885, for example, drew at least four hundred entries.[37] In 1888 Canadian breeders formed the Canadian Kennel Club and set up their own registry. One of the chief instigators (and the club's first president) was Richard Gibson, who had sold out Samuel Campbell's Shorthorn herd at the New York Mills sale of 1873. (Gibson knew only too well how pedigree could affect the value of animals and how the marketing of purebred stock worked.) The breeders who formed the new club said it was created "to promote the breeding of thoroughbred dogs in Canada, to protect the interests of owners as well as breeders, to formulate rules of government of dog shows and competitions, to recommend suitable judges for the same, and to open a proper Register for the registering of dogs in Canada." Pedigrees in such a registry would require the names of the dog, its sire and dam, and its grandsires and granddams.[38] By late 1889 the Canadian Kennel Club had been recognized as autonomous by the American Kennel Club, and the two clubs agreed to uphold each other's rules. Dogs with papers from either could compete in the other's shows without compulsory registration in the other's studbook.[39] The Canadian club's relations with the American club changed over the years. In 1894, for example, the American Kennel Club rescinded the right to show Canadian dogs without American pedigrees, following the policy that American cattle registries generally promoted in that period: enforced registration in American books.

Pedigree issues between the Canadian and American kennel clubs reflected the general international purebred situation. The main restrictions on the international market for dogs developed at the beginning of the twentieth century, growing out of regulations attached to the 1897 Dingley tariff. In 1906 the United States Department of Agriculture, under the authority of the Treasury Department, passed a ruling that all dogs imported from either Canada or Britain had to be registered in the American Kennel Club in order to be imported duty free.[40] When that order was rescinded in 1911 (dogs then could enter as purebred if registered in their place of origin), the *Collie Folio* in Britain noted the loss of income to the American Kennel Club. "That diverts at one fell swoop many thousands of dollars per annum from the AKC exchequer," the *journal* pointed out.[41] British dog breeders saw, just as cattle breeders in Canada and Britain did, that the USDA's 1906 decision had been lucrative for the American-based registry. After 1911 pedigree recognition among the three countries rapidly became reciprocal. In Canada, by 1915 dogs were automatically eligible for the Canadian Kennel Club if registered in either the American or the British Kennel Club.[42] The *Collie Folio* noted that the papers of all three kennel clubs were equally valid, and it gave the pedigrees of all Collies registered each month in any of the three clubs. In 1914 the American Kennel Club and the Kennel Club in Britain made formal agreements about upholding each other's rulings, like those made between the two North American clubs.[43]

War, of course, affected these patterns. It slowed down the transatlantic movement of purebred dogs. It also affected the production of purebred puppies in Britain. To preserve food for humans, the government put pressure on the Kennel Club to reduce the dog population. The Kennel Club decreased pedigree registration during the war, and breeders volunteered to produce fewer puppies. Many dog shows were suspended. In return, the government did not increase taxes on purebred dogs. In the United States, the war provoked a similar disfavor for the large-scale production of purebred dogs. In Canada war seemed to consolidate government control of pedigrees and dog breeding. In 1915 the Canadian Kennel Club came under Canadian law regulating purebred breeding. The club incorporated under the Live Stock Pedigree Act and moved its registry to the Canadian Live Stock Records Corporation. Registration of dogs doubled over the next two years.[44]

The American Kennel Club and the Canadian Kennel Club, in spite of their differences, were both derivatives of the Kennel Club of Britain, and all three shared certain basic characteristics. One fundamental similarity was the structure of their registry systems. Each kennel club ran a single registry for all dogs, with subsections for each member breed, and maintained the right to accept or reject purebred status for every breed it recognized. Breeders therefore had little control over pedigree standards for their particular breed. The idea of breeding up was also foreign to the pedigree system for dogs. Although dogs with unknown background could be admitted to pedigree status in the British registry until 1914 (and in the North American records if they had been allowed pedigree status in the British registry), no formal programs received consideration for breeding up. No mechanism existed to allow registered animals to be used to breed up others to purebred status. The mentality behind the studbooks in all three countries supported a closed system, even though many early entries had no pedigree at all. The pedigree structure for dogs, then, differed from that for cattle in all three countries. Dog breeds did not have separate books, breeders had no say in pedigree standards, and dogs could not be bred up to purebred status even at the beginning of public record keeping.

Individual dog breeds functioned in a similar way under the three kennel clubs. Specialty organizations regulated canine affairs by breed. Various breeds formed their own organizations, called specialty clubs, under the auspices of the kennel clubs, and set the standards of conformation they considered desirable. (If there were too few breeders to do so, the kennel clubs themselves could act as specialty clubs.) The Collie was one of the earliest breeds to have specialty clubs. Breeders in Britain formed the first Collie Club in 1881, which still exists, and regional organizations developed under its head. The British Collie Club wrote the first standards describing desired breed type, and the North American outlines of an ideal Collie were close replicas of the British version. In 1886 fifty people founded the Collie Club of America. Various regional groups came into existence under this parent organization, which regulated shows and published books and magazines on the breed over the years. The Collie Club of America also funds research on diseases specific to the breed, such as Collie eye anomaly. It generally supervises all Collie activities in the country. By 1999 the Collie Club of America had about two thousand members. The present Collie Club of Canada, which acts as a parent association for Canadian breeders, did not exist until 1980, and by 2000 it had about 250 members. Local Canadian organizations for Collies had flourished in the past; for example, an Ontario Collie Club, created in 1884, and a Canadian Collie Club, established in 1897, supported early Collie exhibitions.[45] Collie shows in Canada could be quite large by the beginning of the twentieth century. In 1904 sixty exhibitors with over 250 animals showed under the Canadian Collie Club at Montreal. At the 1906 show in Montreal, one exhibitor brought 35 dogs.[46] Regardless of organizational activities in Canada, Canadian breeders always worked with the Collie Club of America, and of course they continued to be influenced by happenings in the British Collie Club.

The rapid evolution of dog shows and the regulation of dog breeding through these complicated structures soon brought strong criticism from various quarters in Britain, Canada, and the United States. The sport itself—breeding for fancy—did not make much sense to some people. Breeding for fancy, they suspected, had little to do with meaningful improvement, which could be defined as better utility for work, not greater beauty. In 1888 the British *Kennel Gazette* carried an article on the state of Collie breeding, written by a concerned reader. The writer thought that the idea of "fancy" related especially to the breeding of the Collie. Fancies, "locust like, appear to have settled on the Collie," this person wrote, "and unless we can exterminate them, they will most assuredly exterminate the Collie. 'Fanciers' have recently determined that a Collie shall have an enormous head, an enormous coat, and enormous limbs, and that on these three 'points' shall stand or fall the judging; so they have commenced to graft on the breed the jaw of an alligator, the coat of an Angora goat, and the clumsy bone of a St. Bernard." Another person wrote about the specialty club system and promotion of meaningless fancy: "In conclusion, I feel that I must just make mention of the Specialty Club system. . . . Theoretically unmixed blessings, I fear that the clubs, in many cases, have not been for the improvement of their respective breeds. And I am sure that very great caution should be exercised within themselves to prevent them from becoming very hotbeds for the propagation of 'fancy.'"[47]

In 1889 another Collie expert in Britain argued that the system of fancy breeding had done great harm to the naturally good qualities of the breed. "The Collie Club has, by its influence, made our Collies *en masse* more homogeneous," Hugh Dalziel wrote, "and its influence has, on the whole, been to give more correct ideas to the public of what a true Collie is; but it has at the same time allowed the fancy to guide it too much, with the consequence that we have the handsome and useless [model seen in present-day shows]."[48] Writers on show dogs knew nothing about real dogs, Dalziel added. "Thus we have a kennel editor—fortunately of an American paper, though he is an Englishman—telling his readers that it is a peculiarity of the Collie to squint. An English exhibitor, who is a member of the Collie Club, proposed to classify Collies by color, as though they belonged to the same category of animals as fancy mice."[49] Disgust with the show system and specialty clubs increased with time.[50]

In Britain, Judith Lytton presented a rather eloquent commentary in 1911 on the purebred breeding of dogs generally under the kennel club structure. The daughter of Wilfrid Blunt and Lady Anne Blunt (granddaughter of Lord Byron), Judith Blunt Lytton belonged to the landed classes of Britain and later became Lady Wentworth through her mother. Lytton came to be a purebred animal breeder of huge significance. While she bred toy dogs, which drew her into the world of purebred dogs, her greatest contribution to animal breeding was through her work with the Arabian horse. Arabians she produced at the Crabbet Arabian Stud would influence Arabian horse breeding all over the world. She wrote extensively about the breeding of Thoroughbred and Arabian horses and painted them as well.

In her *Toy Dogs and Their Ancestors*, Lytton wrote scathingly about the kennel club system and its effect on the breeding of dogs. Here are a few of her comments about the modern world of dog breeding and showing: "The whole fabric of modern judging is utterly unsound," she wrote. "I . . . maintain that certain types of modern dogs are monstrosities and shall to the end of my days fight against their propagation." "Quality cannot be defined in standards or divided into scales but, like beauty and genius in the human race, it must remain forever independent of legislation." She spoke even more disparagingly of the specialty clubs and their role in judging. "Specialist clubs are, as a rule, merely organs of a few more or less powerful exhibitors; almost invariably dealers of the less reputable kind who have some common aims, but whose interests are not by any means always in the breed they are supposed to present, but alas, in their pockets." "The position of a judge in these specialist clubs is that of a child in leading strings. The judging of some breeds has long been a perfect farce; the dealers play into one another's hands, appoint each other judges and report on their own dogs. Could anything be worse for the improvement of our breeds of dog? The results are disastrous. No wonder respectable people are driven out of the shows."[51]

The connection of shows to meaningless breeding programs also disturbed North Americans. In 1908 Homer Davenport, who was interested in the purebred breeding of animals and best known as an importer of Arabian horses and founder of a registry system devoted to that breed, wrote this in *Country Life in America*: "For a number of years I have been satisfied that the bench shows were destroying the best and most useful breeds of dogs, for the simple reason that the various breeds were scored to a standard made without regard to their useful qualities."[52] James Watson, a dog expert and an early Collie breeder, responded to Davenport's attack on dog shows. Watson pointed out that Davenport had never shown dogs and therefore could not know how fancy affected their usefulness.[53] A beautiful dog, Watson, argued, did not have to be useless. The two characteristics were not mutually exclusive. Dogs could be bred for beauty and work simultaneously.

Concern over the utility/fancy issue led some researchers to test whether breeding for show had altered dogs' ability to work. A study done in the 1920s on the relation of show breeding to usefulness in dogs lent support to Davenport's contention. A small group of scientists and breeders, who suspected that German Shepherds had changed very rapidly after 1890, tried to recreate the breed as it had been before the show system evolved and the breed's characteristics had been altered. These people argued that "the modern German Shepherd had been so carefully bred for 'show points' that, to a great extent, it had lost much of its original value as a utility working dog."[54] After a careful assessment of breeding and showing records, the researchers found that "while certain lines . . . produced champions which qualified for both show and work, such combination occurred only in the *early* history of the breed. The last whelping date for a combined producer [was] 1909. Soon after the breed began to be considered seriously from a show standpoint there came a parting of the ways. One branch turned to the show-ring and the other to the field of work and utility."[55] The researchers drew up a new breed conformation standard, derived by noting the looks of good working dogs that had won championships in the show ring and by studying pedigrees. Modern biologists who study dogs confirm that breeding for beauty points, a form of specialization, undermines other characteristics of a breed that often relate to both temperament and working ability.[56]

While Watson might have argued that showing dogs did not detract from their usefulness, Davenport's remarks hinted at another issue embedded in breeding for show conformation. Shows and the system of specialty clubs promoted the breeding of dogs that could be described as useless or inferior not just because they had been specialized for beauty but also because they lacked authenticity, meaning they deviated from the original type. And divergence from original type could not be defined as improvement; it represented degeneration. The question, in a nutshell, was, How could "improvement" be achieved through breeding for fancy when such breeding led to a new type? Did dogs improved for shows also demonstrate improvement over original type, or could it be said that they represented a lack of quality because of the basic deviation from the older style? As breeds became more standardized to specific conformation standards, a question haunted many breeders. Was quality an issue of closeness to original type, or should that type be molded into newer "improved" forms?

The social respectability of the sport also concerned people from the early days of dog shows. In spite of the elite structure of the Kennel Club, it seems from Lytton's comments quoted earlier that there had been considerable middle-class penetration of British dog showing. Lytton did not favor that development because she believed it reduced the propriety of breeding dogs. The need for greater respectability and more elitism in the

Members of the royal family enjoying their dogs. Queen Alexandra, wife of Edward VII, and her grandchildren visit with dogs of various breeds at the royal
kennels. Note the two Collies, and a third in the distance. In view of the controversy surrounding the Borzoi's relationship to the Collie, it is interesting that
a Borzoi appears between the two of the Collies in the painting. *Queen Alexandra with Her Grandchildren and Her Favourite Pets*, by T. Blinks and F. Morgan, 1902,
oil on canvas, 49 by 64 inches. Collection of Her Majesty the Queen.

sport concerned others besides Lytton. Another British example of the
demand for elitism in dog exhibitions is the formation and functioning
of the Ladies' Kennel Association, founded in 1894, which sought to bring
refinement by establishing dog shows for women. The association, which
joined the Kennel Club in 1899 as a special branch, continued to hold its
own shows until 1907.[57] It published a journal for a short time after 1894,
and there ideas about respectability and dog shows emerged most clearly.
The *Ladies' Kennel Journal* demanded absolute accuracy of pedigrees and
honesty toward buyers. "In conclusion," the *Journal* said in its first issue,
"The Ladies' Kennel Association [hopes], by the example of its Members
at Exhibitions, to purge Shows and the Show-ring of much that is objec-
tionable, and to prove that, though they may be a society of what a certain class of critic is pleased to call 'doggy' women, they are, none the less,

gentlewomen."[58] By 1897 the journal seemed well pleased with progress in that direction. But respectability continued to be related to social class. Reporting on a recent exhibition held by the association, the journal stated: "The rise of the dog in fashionable esteem deserves to be counted as one of the phenomen[a] of the later Victorian era. When the Queen was young, . . . usually [a] disreputable, gin-drinking, prize-fighting ruffian, with a straw in his mouth, a ferret in his pocket, and bull-pup between his legs was the correct type of dog-show frequenter. Yesterday, at the Botanic Gardens, half the Peerage seemed to be present at the Ladies' Kennel show, and the list of prize winners reads like a report of a Queen's drawing room."[59] Elitism remained part of the sport of dog breeding and showing, even if it was intended to be a spectator activity designed to draw middle-class observers. Dog people in North America were also concerned with respectability and elitism and their connection with dog shows.[60]

Although the system for breeding purebred dogs started in Britain, a transatlantic trade in dogs quickly sparked an international industry. Pubic pedigree keeping played an important role in these markets, and British dog breeders had learned their lesson from earlier cattle and horse breeders. A market for dogs bred for "improvement" existed in the New World, but the animals could be defined as improved only if they carried pedigrees. By looking more closely at a certain breed, we can learn more about the internal dynamics of the dog fancy. The next chapter follows what happened to Collies as breeders tried to "improve" them for conformation show purposes, and thus for success in the marketplace.

# Four

# PATTERNS IN COLLIE BREEDING
# AND CULTURE

"Fashion is proverbially fickle; but the fashion which set in more than thirty years ago, and placed the Collie in the exalted position among Domesticated dogs that it [has] enjoyed for so long, assuredly cannot be called by that name. . . . Nor [has] the popularity simply been confined to our own country," wrote a British Collie expert in 1921.[1] He touched on some major issues in the international Collie world: fashion, authenticity, popularity, and markets. All these questions arose from a fundamental driver of that international world: trade between Britain and the United States. Questions about quality, purity, and improvement—and the association of all three with fashion, fancy, or authenticity—built a complex Collie breeding culture within that market-focused world. This chapter looks at the history of Collie breeding in relation to these issues by examining certain topics: the culture of public pedigree keeping; the work of a Collie entrepreneur named William E. Mason; the old-fashioned Collie movement; the association of the Collie name with purity and markets; and attitudes toward Collies as revealed in literature.

What we today know as the Collie is only one breed that sprang up from a base collie type. Early canines of this sort herded sheep in Scotland and northern England for centuries and after 1860 would be bred more selectively in two ways: for utility (herding) and for fancy (beauty points). The Border Collie (developed in Britain) and the Australian Shepherd (developed later in the United States) are utility types that resulted from more organized selective breeding. Selection for fancy led to the modern dog we call the Collie—the breed discussed here. But looking briefly at what the breeders of utility collies sought to do spotlights problems inherent in the fancy Collie world, because over the years utility concerns would never be completely detached from production of the fancy Collie. Breeders of Border Collies and Australian Shepherds tested their dogs' ability to herd sheep in trial competitions that had no connection with the kennel club structure. The first competitions of this nature ex-

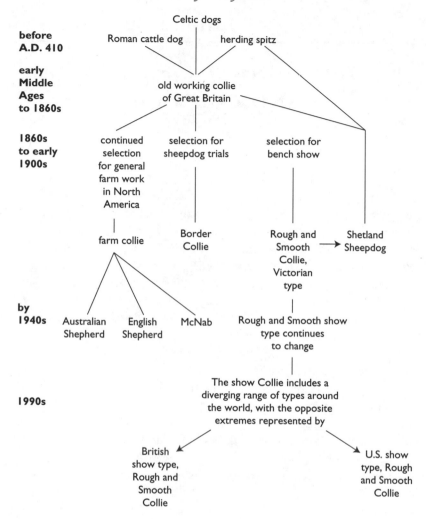

The development of the Collie. The modern show Collie is only one breed that arose from the base type of "collie." This chart shows in a rough way how various branches developed. From "Collie Family Tree," *Friends of the Old Farm Collie,* bulletin 5 (December 1997), izebug.syr.edu/~gsbisco/bull5a.htm (Web site run by gsbisco@mailbox.syr.edu, with permission from Linda Rorem); redrawn by William L. Nelson.

isted before either Border Collies or Australian Shepherds received any "breed" recognition. Trials for canine precursors of modern Border Collies began in 1873 in Wales. Trials for early working collies, similar to those in Britain in the 1870s, started in the United States as early as 1881.[2] Good performance at these trials became the criterion for selective breeding. When the International Sheepdog Trials Society set up rules for competition in 1906, these in effect became the standard for what would become the Border Collie. Formally named about that time because the dogs were found in the border area between England and Scotland, the Border Collie became a breed based on its ability to perform certain tasks, not on its looks. That situation would change when some Border Collies began to be bred for conformation classes in Britain. This shift in breeding motivation caused rifts among British Border Collie breeders.

Border Collies could be found in the United States from the time they appeared in Britain, but an association to support trials—the North American Sheepdog Society—was not founded until 1923.[3] Public record keeping of pedigrees also began for these dogs, but not as part of the American Kennel Club. Breeding for conformation did not become an issue until 1995, when Border Collies started to be shown within the kennel club system. The breed received membership status in the American Kennel Club that year. (Border Collies could not be pedigreed by the Canadian Kennel Club at the time I wrote this book.) The American Border Collie Association, however, refused to recognize dogs based on conformation and stated that animals so bred would not be accepted in its registry until they had passed a performance test.[4] Border Collies, the association argued, could be defined as such only by their ability to work.

Breeding for fancy versus breeding for work plagued the fortunes of the Australian Shepherd as well. Though the breed was well known as a working dog in the United States in the late nineteenth century (it originated there, not in Australia), it was not organized until the middle of the twentieth century. The Australian Shepherd could hold public pedigrees after 1957, through the National Stock Dog Registry. The Australian Shepherd Club of America provided public recording for the breed in 1971. The problem of recording and the relation of registry to utility and fancy, though, soon caused conflict among breeders of Australian Shepherds. By the 1980s some breeders wanted to be part of the kennel club system, but the parent body of the breed, the Australian Shepherd Club of America, remained opposed. Dissident members, who wanted to breed for appearance and to have a registry based on those standards, formed the United States Australian Shepherd Association in 1991. These breeders applied for membership in the American Kennel Club, which was granted in 1993. Public pedigrees could be had under two registries: one based on performance in field or working trials, the other on appearance.[5] Although some dogs held pedigrees in both registries, a considerable number of breeders feared that selecting for conformation would alter fundamental characteristics of the breed. Many breeders of both these utility dogs continued to see the kennel club system, and recording under it, as undesirable because they viewed the fundamental premise the system is based on—breeding for fancy or beauty—as flawed. For breeders of some modern sections of the old collie family tree, it was Davenport's argument all over again: selective breeding should improve dogs' ability to work.

At the same time that selection for herding was being practiced on the base collie type, other breeders bred the same base dog for beauty points. Breeders of the modern show Collie were interested in the origins of that base collie, but even early in selective breeding for beauty they had to admit there was little information on the subject. Hugh Dalziel, the first person to attempt a serious history of the Collie, noted in 1888: "The origin and history of the Scotch Collie as a distinct breed are, to me, unsolved questions."[6] In spite of this lack of information, early lovers of Collies often claimed that the base collie was the original domesticated dog and therefore that of the modern breeds the Collie was most closely related to the wolf. Some believed this had been Charles Darwin's conviction.[7] In the 1880s, an editor of kennel matters in a London journal stated that the Collie bore a close resemblance to the wolf.[8] In 1903 the president of the Collie Club of America claimed that naturalists believed that the Collie had evolved from the oldest breed of dog and was surprisingly wolflike in appearance in earlier times.[9] No evidence supports the idea that the Collie is more closely related to the wolf than any other breed. In fact, modern genetic research suggests that all breeds, not just those that stemmed from the base collie, are equally related to the wolf.[10] The base collie, whatever its background, appeared more commonly in England by the mid-nineteenth century, when trains made it easier for sheep and their attendant dogs to reach cities. The dog of Scotland, which became the modern show Collie, was transformed through the breeding strategies of only a handful of men living in or near industrial English cities. It was Birmingham that grew to be the center of dog breeding in general and of Collie production in particular. The great breeders who in effect created the modern show Collie—A. H. Megson, Charles H. Wheeler, Thomas H. Stretch, R. Tait, Hugo Ainscough, R. H. Lord, William W. Stansfield, and

William E. Mason—all lived near Birmingham or in the area of Lancashire just north of Liverpool.[11] The modern show Collie, then, was a product of England, based on Scottish working dogs. From the earliest days of selective breeding on the base collie, breeders of the conformation show type argued that the name Collie should be applied only to their kind of dog. Even though the word came to signify this breed alone, the idea that it should did not go unchallenged.

Animals exhibited as Collies appeared at the Birmingham dog show in 1861. Many of these early show Collies had been bought cheap from local farmers, butchers, or shepherds by men who believed they were the type that could win in shows. Part of the sport of showing dogs in the early days was the ability to find a beautiful dog, not to breed one. Men who exhibited dogs showed their mettle by finding good dogs in obscure places, buying them cheap, and then campaigning them successfully in the show ring. It made no sense to undermine the hunt by revealing either where they had found the dogs or what their genetic background was. Purity of breeding—freedom from the input of other breed types—played no role in choosing early show dogs that would subsequently be used for breeding. The story of one important early dog, Old Cockie, demonstrates these patterns. Old Cockie, whelped in 1868, came from an unknown background and breeder, but William White bought the dog and showed him at Birmingham with great success until 1873. White steadfastly refused to say where he found Old Cockie. The dog brought £3 from Dean Tomlinson in 1875 when he was sold at the Birmingham Horse Auction.[12] Because the owners of these early dogs took pride in hiding the origins of their purchases, their ancestry could not be known.

Regardless of his lack of lineage, Old Cockie played a fundamental role in early Collie pedigrees. It is the position accredited to the dog in these pedigrees, however, not the mere recording, that is particularly interesting. The evaluation of Old Cockie as a sire reveals important cultural aspects of record keeping. Charlemagne, grandson of Old Cockie, generally gets credit for being the first true modern Collie, but Old Cockie's role as a foundation sire behind Charlemagne goes unrecognized because the line descended through Old Cockie's daughter, not his son. Trefoil (a dog bred by Kennel Club founder Sewallis E. Shirley and whelped in 1873), sire of Charlemagne, holds the title of foundation sire.[13] Family trees in pedigrees traditionally came to be described by analysis of ancestry that passed from male to male—what is called tail male descent. Because Charlemagne's dam, Maude, broke the tail male descent, Old Cockie does not get the title of foundation dog of the breed. The situation is especially fascinating because Old Cockie's huge impact on the breed was well recognized at the time, even if not understood in a genetic sense. He single-handedly changed the dominant color patterns of the breed by introducing sable and white. By 1890 a vast number of Collies came in this color, white and gold to mahogany. In contrast, Trefoil had no effect on such a large scale. Even Old Cockie's enormous contribution to the Collie's development, through the changes he brought about in coat color, could not overcome this break in tail male descent. Not only should breeding be done through selection of males, the thinking went, but pedigrees should be structured around males. The advent of Mendelian genetics did not change either aspect of that gendered breeding tradition. Note from the comments of an American Collie breeder in 1910 how tenaciously breeders clung to the idea of selective breeding through males.

> One doctrine, often practiced and wrongly recommended, is the constant change of sire. It is one of the faults of the present-day methods, and at variance with the practice of former days (as applied to horses, sheep, and cattle), which we should assist in restoring. It is obvious that the accepted strains must be those which descend in direct male line. There are instances in which the dam has been the potential factor in producing high-class offspring; but, even so, they were most likely transmitted to future generations by direct male, because the sires most freely used are those who have come most prominently to the front. At a rough guess, three [times] out of four the sire is the potential factor.[14]

In these circumstances the genetic role of bitches tended to be both lost and underrated.

Collie breeding expanded rapidly after Old Cockie's time, and Collies became popular in the New World. In 1878 the first show Collies entered the United States, and breeders exhibited the dogs at the Westminster Show that year. One dog exported to the United States as early as 1878 cost its new owner $1,000.[15] The late nineteenth-century poultry press revealed the popularity of the Collie within the North American dog fancy at this early stage of the breed's development: "There has been much said and written about the great intelligence and sagacity of the Shepherd dog, particularly the Scotch Collie, and yet his merits are not over estimated in the least," said the *Canadian Poultry Review* in 1878. "We know of no dog which will better please the farmer, stock breeder, poultry fancier, or country gentleman, than a well bred Scotch Collie."[16] In 1885 the *Canadian*

Old Cockie and his grandson, Charlemagne. This portrait of these famous early Collies shows how much the modern show dog had developed by 1880. Charlemagne, the dog on the right, displays distinctly modern features compared with his grandfather. Note in particular the shape of the head and the ears. These changes had occurred over the short span of ten years. *Old Cockie and Charlemagne,* lithograph by Arthur Wardle.

*Breeder and Agricultural Review* quoted the *Ohio Poultry Journal,* which had this to say about "Scotch Collie Dogs":

> The Collie seems to hold full sway in the esteem of all farmers and poulterers, and wisely they have made their choice. A more beautiful and useful member of the canine race cannot be found, and his peerless intelligence and great affection for those with whom he is associated have made him a place in the hearts of all. It is a noticeable fact that when a farmer buys a Collie he never wants to part with him at any price.... The Collie is a general purpose dog.... For all purposes he is *the* dog.... The above assertions may seem "sweeping," but they are warranted by the facts.[17]

A healthy transatlantic Collie industry supported the rise of large kennels in North America. In 1888 the banker and financier J. P. Morgan set up a Collie kennel called Cragston on the Hudson River in New York and paid considerable amounts for British dogs. He purchased Wishaw

Clinker, for example, for $4,000. Samuel Untermyer, who joined the Collie breeding ranks in 1904, competed with Morgan for the ownership of important Collies. Both wanted Squire of Tytton, for example. Untermyer won in this case and purchased Tytton in 1906 for $6,500. Morgan and Untermyer each regularly housed over one hundred Collies. Collie operations tended to be centered in New York and New Jersey, but in 1908 William Ellery, a Californian, paid $10,000 for the famous Anfield Model.[18] Canadians entered the Collie market as well. By 1905 W. Ormiston Roy of Montreal, after buying all the Collies from the Balmoral kennel in Ottawa, housed what some considered the best collection of Collies in the world.[19] Mrs. Gordon of Toronto also kept a large number of imported Collies at her King Edward Kennels.[20] Women could be breeders at large kennels, especially in North America. Clara Lunt's Alstead Kennels, established in 1902, would be based for many years on British imports, and she normally housed at least sixty dogs. By the 1920s Florence Ilch had started Bellhaven Kennels in New Jersey, a stud that produced hundreds of Collies.

A strong market in North America meant that some British breeders could earn good money from their canine operations. As the great American Collie expert O. Prescott Bennett noted, "It may readily be conceived that as a result of the quickening of the pulse, breeders and dealers found the Collie industry a remunerative pursuit, in evidence of which there [were] men . . . whose proud boast [was] that, for many years, their livelihood and those of their wives and families, [had] been entirely derived from that avocation."[21] Breeders found the Collie market very lucrative and took full advantage: breeding Collies became a profession. Under these conditions, successful British breeders also paid high prices for dogs bred by other British breeders (just as high as those paid by American buyers). In 1894 William E. Mason sold Southport Perfection to A. H. Megson for the sterling equivalent of U.S. $5,000. In the late 1890s Megson paid the sterling equivalent of U.S. $8,000 to Thomas H. Stretch for Ormskirk Emerald.[22] One reason the Collie market in North America remained so strong before the 1920s was that in spite of their desire for the animals, people in the United States and Canada could not breed good dogs. Collie after Collie arrived in the New World, but they produced their most important progeny before they left Britain. Various reasons could be offered. Some Collie and dog experts of the time argued that Americans simply found it easier to buy good dogs than to breed them. Probably importing fewer good bitches than good male dogs played a role in the quality in Collies born in North America. Collie breeders also believed that Americans had not yet learned the skill of using high-quality dogs and bitches well in breeding programs—that is, finding bitches and dogs that best complimented each other's good qualities. Good breeders often had an instinct for which animals would best complement another's good qualities and compensate for bad ones, and they usually could not explain how or why. Whatever the reason, modern American breeders generally concur that their compatriots could not breed good Collies before at least 1920.

The continuous draw on British dogs raised some question whether British breeders rigged the market. Judith Lytton, for example, believed that only trash was exported to the United States. "The biggest dealers are not satisfied with getting first-class prices for good dogs," she wrote; "on the contrary, they want to keep their good dogs and win all the prizes themselves palming off the riffraff and misfits on the unsuspecting fairy prince for the fabulous prices that only fairy princes [here she meant Americans] demand."[23] The dogs themselves suggest otherwise. The best Collies Britain produced were exported, but until well into the 1920s they left no good progeny from their breeding careers in North America. Magnet, for example, arrived in the United States in 1921. Though he was often called "sire supreme," Magnet's main impact resulted from the work of his grandson, Eden Emerald, a dog bred in Britain. Magnet himself did not produce worthy Collies in the United States.

One person in particular stands out in the international Collie world because of his unusual ability to breed dogs, market them, and create a demand. His work illustrates the dynamics of a market that supported the breeding of dogs, and it also reveals how trade issues affected breeding principles. Nothing shows the significance of the transatlantic trade to Collie fortunes better than the story of William E. Mason's work. Born in Britain about 1867, Mason entered the Collie world in 1890 when he started his Southport Kennels by buying a seven-month-old puppy that he named Southport Pilot. Mason paid a lot for his first puppy: £100, or the sterling equivalent of U.S. $500. The following year he purchased a renowned dog for £450 and named him Southport Perfection.[24] (As noted earlier, Mason subsequently sold this dog to Megson, another British breeder.) Mason demonstrated at the beginning of his dog career that he could recognize potentially valuable animals, and he would pay a good deal for them. A strong Collie market made such cash outlays reasonable.

Because of his immense Collie knowledge, Mason did not hesitate to buy an animal he believed was of good quality, even if the dog came with an ordinary pedigree from a relatively unimportant breeder. As a result, he did not breed some of the greatest Collies he owned, campaigned, and sold—dogs descended from parents with less than brilliant show careers. Perhaps one of the best examples of Mason's ability to find a valuable dog was the great Southport Sample, Mason's favorite dog of all time. Sample was bred by H. J. Lewis from two dogs said to be of little repute, but Mason took this dog to championship and also marketed his progeny successfully for a short but critical period. Mason's understanding of Collies became so widely known that he began to act as a buying agent and became a broker for top Collies. He sold to both British and American breeders. O. P. Bennett credited Mason more than almost anyone else for the increase in monetary value of Collies over the turn of the century.[25]

His greatest marketing skills, however, emerged in his dealings with Americans. J. P. Morgan and Samuel Untermyer bought Collies through Mason, as did William Ellery. By the end of 1910 Mason had negotiated the sale of at least forty-seven top Collies to the United States. Perhaps the three most famous dogs Mason sold in his lifetime were Squire of Tytton (bred by W. Thomas Horry and bought by Samuel Untermyer), Anfield Model (bred by H. Galt and sold to Ellery), and Model's descendent, Parbold Picador (the dog appeared to have been whelped by Hugo Ainscough because of the kennel name Parbold, but it was bred by W. Preston and purchased O. P. Bennett).[26] Mason crossed the Atlantic at least sixteen times in making his dog deals, and he became a life member of the Collie Club of America. By 1905 Mason's show record had reached incredible heights. In that year he attended forty-eight dog shows and received over three hundred first prizes.

Closely allied with Mason's breeding and marketing career in Collies was his work in journalism. In 1908 he bought the *Collie Folio*, an impressive publication with a wide circulation even though it had begun only two years earlier. Initiated in 1906 by Thomas Baker, a professional dog photographer from Birmingham, the *Collie Folio* remained an unusual magazine for its time. Large, printed on beautiful paper, and containing magnificent photographs of Collies, it became a dog journal deluxe, unlike any other contemporary publication even partially devoted to dogs. The very existence of the journal demonstrates just how strong the interest in Collies had become by the beginning of the twentieth century. "In submitting this, the first number of 'The Collie Folio,'" Baker wrote,

> the Editor would express the hope that it will fill a void long apparent in the Collie world. The Collie fancy being so world-wide and influential gives him hope that this humble effort to fill the void will be successful and [win] acceptance [by] the numerous lovers of

the Collie, a dog that deservedly takes the very highest place in the canine family, not just in monetary value, but in intelligence also. Signs are not wanting to prove the wonderful strides the breed recently made, and also its increasing popularity. From America, South Africa, and our Colonies, come the glad tidings of a desire and a determination to emulate and excel, if possible, the success of the old country.[27]

As later issues of the *Collie Folio* made clear, the "world-wide" influence of Collies was largely restricted to the Anglo world, namely the British Empire and the United States. Collies did not become particularly common in European countries at this time, although some could be found in Germany. It might be that the close relationship between British and German royalty in Victorian times, and Queen Victoria's favoring of the Collie, helped establish the breed in Germany. In 1889 Collies could receive public recording in that country.[28] By 1907 the *Collie Folio* reported on the activities of eighteen Collie clubs: eleven in Britain, four in America, one in New Zealand, and two in Germany.[29]

The journal started when the major early breeders were still alive and doing business, so Baker could draw on the best information available about Collie origins and the foundation of Collie breeding culture. That advantage makes the *Collie Folio* invaluable for facts about early Collies, their breeding, and the ideology of the men who created the breed. One of the first issues carried information on the history of the Collie by Charles H. Wheeler, a man involved with Collies since the 1870s. Wheeler often addressed his remarks to new Collie fans, and while he explained certain aspects of early Collie breeding to them, he also touted ideas that had not been part of that world. For example, Wheeler warned that Collie greatness, could be preserved only through maintaining purity of the breed, and by this he meant there should be no introduction of foreign genetics. "It is my ardent desire," Wheeler wrote, "to do all in my power to serve the best interests of the Collie, to keep him pure and uncontaminated."[30] The breeding background of dogs like Old Cockie (which various writers discussed at times in the journal) made it clear that purity did not concern original breeders like Wheeler. From the beginning, Baker carefully included news from the United States, the main marketplace for Collies. The *Collie Folio* not only reported on breeding and showing in the United States, it also followed American Collie gossip. For example, an argument between Alfred Blewitt, the English kennel manager of J. P. Morgan and Alvin Untermyer, son of the breeder Samuel Untermyer, about the merits of the dog Wishaw Clinker received careful attention, as did the show career of Squire of Tytton.[31]

In January 1908 Mason brought out his first issue of the *Collie Folio,* and changes in the journal immediately became apparent. One new aspect was the emphatic marketing of Mason's own dogs, and that pattern illustrates his mastery of advertising. On the cover of the first Mason issue, and many issues to follow, was a picture of Southport Sample. Mason also promoted the Southport kennels in large, skillfully focused advertisements. He concentrated on particular dogs and their progeny, placing great emphasis on his favorite, Southport Sample. In 1909, for example, on a full page describing Sample's progeny, Mason claimed the dog had sired nearly one hundred good Collies.[32] By January 1910 Sample was touted as the winner of 136 first prizes and as the sire of one hundred winners in England, North America, and Europe. Advertisements also stated that his progeny had brought higher prices than that of any other dog.[33] Because even at this time the Sample type did not dominate the show ring, one wonders about the truth of some of these statements. Other changes Mason made were to enlarge the journal and to include bigger sections devoted to American breeders. Mason knew where his real market lay. Reports by judges of shows in Britain and North America were also featured more prominently. Every Christmas issue contained a large advertising section with layouts on various breeders in both North America and Britain, and Mason himself wrote commentaries on their breeding programs. The size of kennels was shown by the number of upcoming puppies advertised. Kennels often stated that they had as many as sixty a year available.[34] Commentaries and reports on Collie happenings in the United States, Canada, and New Zealand meant that the international world of Collie culture could be found in the *Collie Folio.*

Mason cemented his relations with American breeders even further when in 1909 he established a kennel at Ridgewood, New Jersey, in partnership with William Ellery of California. With Ellery's contacts in California, Mason could say he had access to markets across the United States. But his ability to sell Collies and his new American breeding operations interfered with his role as Collie expert, exhibitor, and judge in the United States. American breeders argued that his breeding and selling created a conflict of interest with both his exhibiting and his judging. Mason bowed to public opinion. In his January 1911 advertisement in the *Collie Folio,* he announced his intention to sell his dogs at Ridgeway. But he claimed he had not operated in any dishonorable way. He stated:

It has been presented to me that, having sold practically every Collie of good repute for importation into this country during the last ten years, I ought not to exhibit my present stock. I have shown but five times during this past year. Surely not an unreasonable program. On each of these five occasions I have incurred the displeasure of the owner of some former inmate of my kennels. I consider my very modest exhibition essays quite justifiable, but have decided to bow to the wishes of my clientele. I therefore offer my present stock for sale at prices that will defy competition, quality considered. On the completion of such sale I shall retire absolutely from the American show ring.[35]

He does not seem to have done so, though, because his May 1912 advertisement listed his wins at the Collie Club show in New York for that year, with Southport Sample being the biggest winner.[36] In 1913 he advertised Parbold Picador full out, apparently to enhance a future sale.[37] A sale to O. P. Bennett did take place before the end of the year.

In 1914 Mason sold the *Collie Folio* to William W. Stansfield, the breeder of Laund Collies. The journal did not manage to maintain the excellence it had under Mason, and it survived only until 1917. War, of course, interfered with its ability to maintain wide circulation. The general restriction of purebred dog breeding, the reduction of showing, and the virtual shutdown of the transatlantic trade in Collies shrank the potential number of readers. Deliberate changes by Stansfield, however, could be felt beyond the effects of war. Between 1914 and 1917, Stansfield slowly replaced the large front cover pictures of Sample and Southport with pictures of Collies from Laund Kennels. Also under Stansfield, the input of American breeders received more attention, especially the section "Prattings from U.S.A.," generally written by Eileen Moretta, a well-known actress and Collie breeder and also the importer of the great Magnet. Stansfield, however, lacked Mason's promotional skills.

Meanwhile Mason's Collie fortunes began to wind down. The collapse of the Collie market caused by the war no doubt played a role. He had decided to sell the journal because of his new commitment in San Francisco, where he had agreed to run the 1915 dog exhibition at the Panama-Pacific International Exposition. There did not seem to be much for him to do in the United States after he finished his work there, so he returned to England during the war and took up his dog breeding, but in a much reduced way. In 1938, by court order owing to complaints about noise, the dogs remaining in the kennel were dispersed. Mason died in 1941.[38]

Over his life's work with Collies, Mason would be involved in many breeding controversies. One was the old-fashioned Collie movement, which could be defined as the desire to return to, or at least preserve, the original collie type, bred for utility, not fancy points. Breeding strategies used as early as the 1860s to create the show Collie had centered on the shape of the head. As the first issue of the *Collie Folio* put it, the head "has always been, and justly so, the leading and most important feature in a Collie. A perfect head is a very beautiful thing to contemplate. The perfect structure of the skull is... one of the most difficult problems breeders have to solve."[39] As a result of these breeding strategies, head property conflicts would be pivotal to the rise of the old-fashioned Collie movement. Three specific factors lay at the heart of the matter and enflamed the debate.

First, many people came to link the narrow heads that breeders sought in the fancy Collie with stupid and even vicious dogs. Complaints about Collie heads emerged almost as early as breeding for fancy did. In 1877, for example, a Collie writer noted that some breeders wanted "tapering jaws as long as those of a pike."[40] By the late 1880s at least one Collie expert believed this breeding strategy limited the intelligence of the bench show dog.[41] By 1900 the Collie head had changed enormously from what it had been like in the late 1880s, and it was by then commonly thought that modern show Collies lacked intelligence. The *Farmer's Advocate* in Canada noted in 1906, that "the yellow, long-nosed, narrow-browed 'fancy' dog... is not a competitor in the lists of intelligence."[42] *Field and Fancy* in the United States thought the same thing and also reported that year that modern show Collies had narrow heads and therefore no brains.[43] Breeders staunchly defended their breeding policies. The president of the Collie Club of America argued in 1903 that breeders did not sacrifice brains for show beauty in Collies.[44] The *Collie Folio* agreed. "Some critics argue that the modern Show Collie is deficient in brains" because of lack of room in the head, the *Collie Folio* reported, then scoffed: "People might just as well contend that the man who wears the largest hat has the most ability."[45]

By the early twentieth century, complaints that head shape in Collies related to intelligence were based to some degree on reality. Attempts to create ever more beautiful dogs through certain breeding strategies had led breeders into serious difficulties. Identifiable breeding programs had created the problem, which could be related to the use of one particular animal. Collie people hailed the advent of Anfield Model (whelped in 1902) because they believed the dog embodied the elusive, sought-after head appearance. A critical difficulty arose, however, when it became apparent that, while in Model and his line breeders had managed to create a pleasing new type with a beautiful head, distinctive, unpleasant temperamental characteristics went along with that look. Model happened to be stupid and even vicious, qualities inherited by many of his most important relatives and direct descendents. Ormskirk Foxall, Weardale Lord (both relatives of Model), and especially Model's grandson, Parbold Picador all carried that trait to a marked degree. Weardale Lord was particularly savage. Breeding for more desired traits found in the Model line could be had, breeders soon found out, only by acquiring the savagery of temperament that Model and his relatives transmitted along with their beauty points.

Many important Collie breeders accepted that savagery, and they continued to inbreed intensely to Model or his relatives in order to perpetuate certain looks. Mason admired, bought, and campaigned both Parbold Picador and Anfield Model. O. P. Bennett, the great expert on Collies who imported Parbold Picador and used both him and his lines extensively, did not seem to consider temperament important in breeding dogs. William Ellery bought the famous Anfield Model himself for $10,000. Model became the supreme star of Ellery's kennel. Model's head, in fact, was for years touted as embodying perfection. The difficulty breeders had in detaching temperament from the overall type they sought, or breeders' lack of concern for the temperamental baggage accompanying certain phenotypic characteristics, made some people question how the conception of "improvement" could be linked to breeding for beauty in the Collie. The connection of savagery to the newer variety as demonstrated by Model and his main descendants left a deep impression. As the Collie diverged further and further from the original type at the same time that show dogs demonstrably became increasingly vicious, polarized views developed about the value of the new Collie and its head shape versus the original collie. Emphasis on breeding for fancy head shapes with no regard for temperament fueled disagreements over what improvement meant, what role authenticity to original type played in quality, and the value of breeding for work. Perhaps the original base collie had been a better dog.

The second factor that played into the old-fashioned Collie movement was the way the dynamics of the Collie trade related to the modern show Collie's head. When wealth dovetailed with the production of modernized Collies that concurrently demonstrated poor temperament, some people argued that money and monopoly underlay breeding programs that emphasized head qualities for fancy points. The degeneration of the Collie might be brought about by the ability of a few rich men to buy particular dogs, but the appetite of the wealthy seemed to be whetted by the system of bench shows. Americans like Morgan and Untermyer, and the competition between them for the best Collie, seemed very attractive to English breeders because of potential sales. Success at English bench shows made these rich American men want the dogs. So, it was argued, desired style resulted from the dictates of the shows, which demanded narrow heads. Both the style issue and its relation to quality and also the monopoly problem bothered many Collie fans.

For example, early in 1908 Homer Davenport, the commentator on dog shows mentioned earlier, who was also an importer and breeder of Arabian horses, linked head properties and stupidity in Collies to the buying habits of Morgan and Untermyer. "The prize-winning Collie of to-day is about as useless an animal as could be imagined. Their heads are long, narrow, and flat, with no room for the brains the breed once had," Davenport wrote. "With this lack of brains they have become suspicious creatures, ready to snap at any moment." Davenport went on to explain how the situation had arisen. "The cause of all this is possibly due to the fact that the men who arranged the standard have been men with little use for dogs, other than for leading them on the Avenue, and year after year they demanded longer and flatter heads for the Collie dog, with less above the eye and more below. That kind of head brought big prices, and the Morgans and the Untermyers paid as much for one of these stupid Collies as a home would be worth on Manhattan Island." "The judge at the dog show," Davenport pointed out, "will tell you that the modern Collie, with so little brain that he has to be led, will bring $10,000, while the old-fashioned Collie, that would go five miles away and fetch home the sheep and cows, would bring from $25 to $500."[46] Collie breeder James Watson responded to Davenport's general attacks on modern Collies by saying, as breeders had said before, that a narrower head did not mean a stupider dog.[47] But Davenport believed, as did others, that Morgan bought dogs as investments and had no love for or understanding of the breed. Other fanciers feared that somehow the two wealthy American buyers had cornered the Collie market for genuinely outstanding dogs. One writer to the *Collie Folio* in 1907, for example, worried more about the effects of Morgan and Untermyer on the size of the Collie market than on the breeding quality of Collies. This fancier claimed that the expensive tastes of a few prominent breeders in the United States damaged Collie breeding because their buying shut out other potential Collie fanciers who wished to compete.[48]

Since J. P. Morgan in particular received much of the blame for perpetuating these problems, anything that can be learned about this man's Collie breeding should be helpful to this story.[49] No records of his Cragston Kennels survived Morgan's death, so we cannot know a great deal about his breeding strategies or his personal involvement in the operation. The little material that does exist on these subjects is interesting, though, because it suggests that some of these popular convictions about the wealthy financier and his Collie involvement were not true. Morgan (who was born in 1837 and died in 1913) became interested in Collies in 1888 when he went to the Westminster show and bought the winning dog, Bendigo, and his dam, Bertha. From that point Morgan rapidly expanded his Collie concerns at his Cragston Kennels on the Hudson River in New York, and he gave every indication that he was personally involved. In 1892 he hired a New York firm to design a kennel. After reviewing the plans, Morgan authorized construction: "The plans & basis are satisfactory to me & you will please proceed with the work with all due dispatch," he wrote to Mead and Taft.[50] The resulting kennels were magnificent, and the *Collie Folio* wrote glowingly about them. Lighted with electricity and heated by steam, they also had beautiful enameled baths for the dogs.[51] In 1903 Morgan hired Alfred Blewitt, a friend of Mason's from England, to run the operation.[52] Blewitt lived on the property and oversaw all show and breeding operations, but there is no indication that Morgan himself did not play a role in making important decisions. Morgan bought the best Collies he could find in Britain, and Mason frequently acted as an agent in these purchases. Morgan housed many famous imported dogs, including Sefton Hero, Wishaw Clinker, and Parbold Purity, and in the 1890s he regularly benched about sixteen dogs at the Westminster shows. He became a life member of the Collie Club of America and served as both president and vice president several times between 1899 and 1908,[53] demonstrating his genuine participation in Collie culture.

From the beginning, Morgan appeared to take an interest in his dogs.[54] In 1893, for example, he refused to sell Bendigo to a prominent Philadelphia breeder. "I do not care to part with Bendigo at any price," Morgan wrote to Henry Jarrett.[55] When Morgan exhibited at the Westminster show that year, the *New York Times* reported on his excitement at the show. "He stood there, a rather impatient man of enormous affairs, much more excited over the possible action of the English judge with the red necktie and the doggy air than he appears when vast financial interests are being manipulated by him. He patted his champion dog, 'Sefton Hero,' on the head [and] talked to him as he would talk to a child."[56] Morgan loved Sefton Hero, and the dog slept under his bed. Many believed that the financier left Collie breeding in 1908 in part because he could not find or breed another dog as good to replace Hero. The dog was mentioned by name in Morgan's obituary.[57] He bought the grand old Collie Southport

Perfection for an unknown price from A. H. Megson, who had earlier paid £1,000 (the equivalent of U.S. $5,000). Subsequently Morgan quietly kept the animal as a house companion at his English Roehampton estate.[58] The Collie world had assumed that Morgan had sent the dog to Cragston Kennels for breeding, but the financier evidently purchased him solely as a pet. Collies appear to have been a part of family life at the Morgan country estate at Highland Falls as well. Photos of the children of Louisa Morgan Satterlee, Morgan's daughter, regularly featured Collies and Collie puppies, and Louisa's husband even painted a watercolor of Hero. "Mr. Morgan obtains much personal pleasure from his Collies, apart altogether from the exhibition side of the matter," stated the *Collie Folio.*[59]

By 1907 rumors circulated that Morgan intended to retire from Collie breeding and that he had given Blewitt all the dogs, valued at £50,000. Some believed one reason Morgan decided to sell was that his wife got tired of the barking of over a hundred dogs.[60] Others contended that Morgan saw that his buying the best Collies hindered the efforts of other American breeders.[61] Perhaps he had been influenced by the sentiment in some Collie circles about rich men's cornering the market. By June 1908, not only was Morgan out of Collies, but Blewitt had sold all the Cragston dogs and returned to England.[62] "It will be regretted by all fanciers in every part of the world if this keen fancier and good judge of the breed decides to give them up," the *Collie Folio* predicted.[63]

Curiously, in light of the attacks made on Morgan about his role in the breeding of thin-nosed, stupid, nonutility Collies, after his death various people wrote to his son inquiring about the wonderful dogs Morgan had owned and hoping to find in these lines an old-fashioned utility collie. The writers usually included a picture of their beloved dogs when they wrote to J. P. Morgan Jr. Wishaw Clinker, a dog of the 1890s, received the most attention in these letters. (Morgan had sold the dog to M. C. Richards of New York in 1908.)[64] In 1918, for example, when Robert Preston wrote asking for a puppy of the Wishaw Clinker line to replace his beloved dog Chan, who had descended from Clinker, he claimed that Chan's great qualities "came down through your father's dog [Clinker] a blunter nose which is characteristic of what is called the 'Old Fashioned Collie,' now nearly extinct. These dogs had little or none of the Spitz blood in them and lacked the treacherous and uncertain temper of the modern sharp nose Collie in addition to being more intelligent."[65] (J. P. Morgan Jr. made it clear in his response that records of the Collies no longer existed and that the Morgan family did not know what had become of the dogs.)[66] Preston connected viciousness and impurity with the modern

Mabel Satterlee and a Collie puppy in 1906. Mabel was the granddaughter of J. P. Morgan, and as the family albums demonstrate, she enjoyed playing with Morgan's expensive dogs and their valuable offspring. Images like this one suggest that the rich fancier did not buy and breed dogs solely for monetary value. From Lousia and Herbert Satterlee's album "Miscellaneous 1904," Archives of the Pierpont Morgan Library, New York.

Collie, and he hoped to return to the old-fashioned Collie through lines bred by Morgan. Morgan might have paid large sums of money for Collies, but writers like Preston make it evident that the financier did not always purchase or breed the extreme type of dog, so that his buying habits did not necessarily fuel the breeding of sharp-nosed dogs.

Preston's comments on Clinker and the modern show collie suggest a third factor in the old-fashioned Collie movement: the speculation that the introduction of Borzoi (Russian Wolfhound) blood (Preston identified it as Spitz) into the Collie had led to the thin head. In other words, some Collies of the extreme fancy show type, with respect especially to head qualities, could be said to be basically impure. Fancy traits had been achieved by infusing blood foreign to the Collie breed. The true, or authentic, representative of the Collie was the older, original version. In fact a Greyhound type of Collie resembling the Borzoi could be found in the ranks of the breed quite early. In 1885, for example, the *Canadian Breeder and Agricultural Review* warned breeders to avoid the greyhound type.[67] The *Kennel Gazette* in Canada claimed in 1889 that at every show of Collies one could see a setter, greyhound, or even Saint Bernard type.[68] It is not known whether this greyhound style resulted from the introduction of foreign blood. A Collie expert writing in the 1940s stated emphatically that Borzoi blood had been introduced into the Collie breed: "There can be no doubt that long after the Studbook was established and pedigree records carefully kept Borzoi blood was surreptitiously introduced into the Collie breed."[69] Curiously, one of the earliest writers on the Collie stated in 1890 that the modern Collie had resulted from an infusion of Scottish Deerhound blood.[70] This breed bore many similarities to the Borzoi, and Deerhound blood could have accounted for what were seen as Borzoi characteristics. Yet later Collie experts did not speak of a Deerhound cross; the Borzoi question haunted the past of the Collie. As late as 1999 Collie breeders still discussed the Borzoi issue, though they claimed it remained only a rumor that such a cross went into the present Collie makeup.[71] The persistent evidence in some late nineteenth-century fancy Collies suggesting such a cross, however, brought the issue of impurity in Collies to a climax. Borzoi blood seemed to be very much part of the extreme show type at the turn of the century, and in the *Collie Folio* Mason fueled the idea that the speculation held some kernel of truth.

Mason may have campaigned modern-style dogs like Picador and Model, but his favorite type continued to be the more conservative one seen in Sample. Perhaps his natural leaning to that kind made him particularly angry because some fancy Collies lacked even the moderation of Model's conformation. Mason believed a number of show Collies had become so extreme that they no longer resembled desirable Collie style. He defined what he objected to in these dogs as their tendency toward Borzoi characteristics. In 1910 Mason decided to campaign openly against this type of Collie.

We need not split hairs, but we do feel, after mature and very earnest thought, that the time has come when a firm stand must be taken against the prominence extended to dogs owning Borzoi characteristics. . . . We do not propose to disgrace the usurpers as Barzois [*sic*], but shall henceforth refer to them as of foreign origin. The word Barzoi may not truly describe the origin of the evil. It is a fact that our biggest and most successful breeder has in the past used a Barzoi in seeking for his ideals. A continual striving after length and fineness of head has evolved many specimens with heads of a totally different type to that of the real Collie, even in strains of pure origin. . . . A cankerous and malignant growth has unquestionably sprung from vigorous roots, and strong measures will be required to stay its further progress.[72]

These remarks brought an immediate response. The next issue of the *Collie Folio* carried a number of breeders' comments. One claimed that the Borzoi type was indeed deplorable, but that it did not necessarily result from a foreign cross of blood. Breeders often, for example, took no interest in the females they used and looked only to male champions to breed champions. Or breeders had become obsessed with pedigrees. "This kind of breeding is, in my opinion, more to blame for the production of wrong type than is the introduction of foreign blood, Borzoi or any other name you like to give it." Another breeder wrote: "Your promised crusade against the foreign element in Collies is not by years too soon." For the past ten years true Collie type had taken a backseat in the show ring. "The disgust of the general public at the utter lack of Collie character or appearance of average intelligence of our present-day show specimens," this breeder added, had ruined the popularity of the Collie.[73]

While breeders heartily agreed with Mason's stand against the Borzoi-type head, they did not support the idea that there had been actual infusion of foreign blood, because they feared they might be labeled as the culprits who had introduced it. Hugo Ainscough, for example, wrote, "In case any of your readers might imagine that I am the culprit, I am writing to say that during my 30 years' experience in Collie-breeding I have never used a Borzoi, neither have I ever used one with any Borzoi cross in it, to my knowledge. I may also state that I have no sympathy with the Borzoi type, and I shall do all in my power to suppress it." Mason responded by saying that of course he did not accuse Ainscough. "The incident occurred some years ago, and we did not think it necessary or advisable to publish the real culprit's name."[74] Breeders became increasingly concerned about statements concerning a Borzoi cross on Collies because it threatened to undermine all their work and their markets as well. One breeder wrote: "I am convinced at the present time there are far less of the Borzoi type exhibited than were a few years ago. . . . You make rather sweeping about a large breeder having introduced the Borzoi cross; so I take this opportunity of stating that none of this alien blood has ever entered the Ormskirk Kennels with my knowledge." Another breeder denied that any Borzoi crossing had taken place by any prominent breeder. This person admitted, though, that a well-known breeder had in his youth experimented with Borzoi blood but had got all of it out of his kennel.[75]

The faint evidence that Borzoi blood had been experimented with and then abandoned appeared in letters from other breeders as well.[76] It seems that any Borzoi blood had been abandoned in breeding programs, so no proof exists in primary sources that it went into the makeup of the present-day Collie. From the work of dog experts it seems that modern DNA testing cannot help us solve the problem with finality. No sequence of alleles that could act as a genetic marker for various breeds has been found to date. Furthermore, studies show that head shape cannot serve as an indicator of genetic relatedness between breeds.[77] Whether Borzoi blood did or did not play a role in forming the Collie head and general looks, however, does not matter to this discussion. The speculation that it did was enough to fuel the old-fashioned Collie movement.

In these head property conflicts a number of convictions came together to stimulate a formalized movement to save the older type: poor mental qualities in top show dogs (Anfield Model and Parbold Picador, for example) that were favored extensively for breeding; wealth as a perceived driver for the creation of narrow heads that in turn meant stupid Collies (seen in the buying practices of Morgan and Untermyer as well as the entrepreneurial work of men like Mason); and the possible infusion of Borzoi blood to accomplish bench show type all suggested that the modern Collie represented deterioration of the true breed. The old-fashioned Collie movement began in 1911, formally triggered by two letters to the editor of *Country Life in America*. One spoke for the old-fashioned Collie, and one spoke for the modern dogs. "As long as I can remember," Otis Barnum wrote, "I have heard stories of the wonderful sagacity and faithfulness of Scotch collies, but somehow, since the advent of the modern, sharp-nosed, show type of collies, these stories have been getting fewer. . . . I wish your magazine could do something to save this dog from extinction."[78] A Collie breeder responded to Barnum as follows: "It is coming to be the fashion for the pessimistic to deplore the passing of the old-fashioned collie and to bemoan the waning intelligence and usefulness of the present-day show type. 'Treacherous,' 'sharp-nosed,' 'automobile-chasing,' are some of the epithets applied to him by his detractors." This breeder went on to say that a separate type of dog had actually developed. "While it is true the present-day collie is different from the old-fashioned type, that does not necessarily imply that he is a degenerate and inferior in every way." "The majority of people in this country . . . have no use for a working dog. They want a dog that is handsome as well as intelligent, and there can be no question of the collie's superlative claims to beauty. He is undoubtedly one of the handsomest, as well as most aristocratic of all our dogs." The writer then added pictures of dogs owned by Untermyer, including one of Squire of Tytton, clearly linking show Collies with wealthy buyers. This writer then described a rescue performed by Morgan's Sefton Hero (and added what Morgan had paid for the dog). "Certainly the old-time collie could not have done better than that."[79]

Portrait of Collies in the 1890s. This painting shows what the "old-fashioned" Collie looked like. Note the thicker head, the heavier ears, and the shorter fur coat than would be seen in a show Collie. Note also the different color patterns. One dog, in fact, is bicolored but with no white. The painting projects the sense of utility through the dogs' association with the sheep in the background. *Collies*, by Wright Barker, 1897, oil on canvas. Sara Davenport Fine Paintings.

The editor of *Country Life* found the problem posed by these two writers interesting and decided to take up the cause of the old-fashioned Collie. "There is nothing to be gained by discussing the question of superiority between the old and new, because they have become totally different," he wrote in 1912. And the new show Collie was definitely here to stay. A separately recognized breed or type, the old-fashioned Collie, needed recognition. "We therefore respectfully bring this suggestion to the attention of the authorities of the American Kennel Club, the Collie Club, the Westminster Kennel Club, and any others whose influence is needed to save the old-fashioned collie."[80] Responses came in quickly. The secretary of the American Kennel Club replied, "I do not understand what you mean by the old-fashioned collie, as there is no such breed." The editor of *International Dogs* said, "Personally I cannot identify the dog you refer to." Mason wrote to *Country Life* as follows. "I am not clear as to what you mean by the Old-fashioned Collie. If it is that you wish to cultivate the sort that was shown twenty or even less years ago, with thick, coarse heads, light eyes, pendulous ears, and vacant expression, then I am quite sure you will never make any headway and I shall do all I can to show the absurdity of your scheme." The editor of *Country Life* claimed that comments like these simply missed the point. The old-fashioned Collie continued to be beloved by many, and many also stoutly believed in that type's superiority.[81] The journal might have noted (but did not) the undercurrent of sentiment in these comments, which suggested that the name Collie should be attached only to the newly evolving breed of show Collie.

Mason decided to pursue the issue in more depth in the *Collie Folio.* "From time to time one hears the wailings and croakings of that portion of the public which believes that the modern Collie is all wrong, and that great retrogression has been the outcome of the labors of those who have striven to improve the grandest of all the canine species. We are amongst those who very much think otherwise." Mason noted the old-fashioned Collie campaign newly initiated by the editor of *Country Life* with disgust and offered these comments: "Every once in a while someone pops up with a wail and a plea for 'the old-fashioned Collie.' Usually such men are they who have owned stock of the type prevalent in the 70's and have done nothing since to improve the points of the breed." But for a reputable journal to take on the cause was shocking. Pictures that accompany *Country Life*'s description of the old-fashioned Collie made "those with an eye for the beautiful ... shudder." Furthermore, he claimed, the dogs pictured looked very much like Saint Bernards or Newfoundlands, indicating that they might be less pure than the show Collie. The bottom line, Mason said, was that a modern fancy Collie could do anything an old-fashioned Collie could. It was just a matter of training.[82]

A number of prominent breeders of show Collies in the United States responded by deciding to test utility in the modern show Collie. They formed the National Collie Association in 1925 and planned to support stock dog trials for show Collies. The organization broke up quickly, however, when one of the founders died, so the movement to test utility with fancy Collies did not last long enough to have any long-term effect. By the 1930s, the old-fashioned Collie movement began to die down, even though the utility/fancy issue remained unresolved, partially because savagery in show dogs had become less strident. Correction of the problem would, ironically, emanate from the Model line, through two descendants of Parbold Picador. The crossing of descendants of two dogs whelped in 1912, Laund Limit (an animal bred by Stansfield) and Magnet, ultimately showed that better-tempered and intelligent dogs could also have desirable show qualities. Interestingly, on his dam's side Limit was closely related through inbreeding to the great Southport Sample, a dog known for his excellent temperament as well as great beauty. (O. P. Bennett described the dog as one of the ten greatest male Collies that ever lived.)[83] Sample does not get credit for this temperament change, however, because his contribution came through his daughter.

Old-fashioned Collies continued to be seen and valued on farms in North America until the 1950s, and some of these held AKC pedigrees designating them as purebred. By the 1980s it had become increasingly difficult to find dogs of this style, but admiration for the old-fashioned type had not disappeared. Lovers of the "old fashioned farm Collie," sometimes called the old farm Collie, began a search for those that remained. In 1995 a group of people who had loved these dogs in their childhood, from both the United States and Canada, formed a group called Friends of the Old Farm Collie. They set out not only to preserve the style but also to establish the old type of collie as a distinct breed. The group planned to start breeding operations either by breeding from surviving examples of the dogs that they found on farms or else by recreating the old farm collie from a variety of breeds. Such breeders had a clear vision in mind of what they wanted, and that was the dog that Collie entrepreneurs like Mason had scorned in 1912. The old-fashioned Collie had distinctive head features—large eyes, ears not too pricked and off the head, and a heavier and wider skull with an evident and pronounced "stop" (where the head domes on dogs like setters). The dog had shorter legs and a shorter coat. In contrast, the show Collie has smaller eyes, a narrow, tapered head with almost no "stop," and ears with a higher prick and set well up on the head. The legs are longer, as is the body, and the coat is much longer.

The question was how the old-fashioned collie could be created with the genetic material available. Should only AKC or CKC registered dogs of the desired type be used as the foundation dogs for a new distinct breed? Or could pedigreed Collies be crossed on dogs such as Australian Shepherds or English Shepherds, or even Border Collies to recreate the old-fashioned Collie? Breeding standards within either framework stimulated discussion. Fans of the old collie type began publishing a bulletin on the Internet that recorded their progress.[84] The search for registered Collies of the old sort turned up a few dogs. One bitch pedigreed by the CKC was found in Ontario. A kennel that bred for old qualities (the operation had been established on foundation stock found in Tennessee and North Carolina) was said to exist in South Carolina.[85] Fans became excited when they learned that an AKC registered black-and-white Collie had been found in Texas. Lovers of the old kind noted that this dog had no tan markings, thereby making him a throwback to the 1860s before the tricolor period.[86] The attempt to recreate the old-fashioned Collie also led to the breeding of Australian Shepherds on AKC registered Collies.[87] But through both processes, people found it hard to agree on what standards should define an old-fashioned Collie, and in 1997 the movement took a new direction. Some members of Friends of the Old Farm Collie decided that their ideal for an old-fashioned Collie closely resembled the early bench show Collies of the 1880s, and they formed the Classic Victorian Collie Club.

> We believe that the old scotch collie physical type range is recoverable from the Collie gene pool. We would like to form a club with the purpose of reestablishing and preserving this physical type, in otherwise healthy, sound Collies.... We would like to work with current Collie breeders, worldwide, who share our concern about health, soundness, function and character and who may be able to help our members find quality Collies that exhibit some or all the old scotch features."[88]

By 2001 the move to re-create the old type of collie had gathered momentum. For example, breeding operations based on the combination of the show Collie and the Australian Shepherd flourished.[89] The American Working Farmcollie Association had also been established with a registration system, making the old-fashioned Collie a purebred breed even if the American Kennel Club does not accept it as such.[90]

The original old-fashioned Collie movement of 1911–12 revealed another sentiment held by many breeders of show Collies. They refused to associate the name Collie with other collie types and clung tenaciously to the idea that the name belonged to their version and theirs alone. That conception had been challenged before 1911 by another breed. Shetland Sheepdog breeders had fought for the use of the Collie name and lost. At a general meeting of the British Collie Club in 1908, Mason brought forward a proposal "that the Club take cognizance of the use of the word 'Collie,' in relation to a breed of mongrels being shown in Scotland as Shetland Collies. It was clear that many of the members present had not realized that the persons responsible for the foisting of this despicable cross-bred toy on the public were gradually inculcating the erroneous idea into the minds of a lethargic public that they had some pretence or right to be dubbed Collies." The club resolved that the name Collie should not be used for the small Shetland dog and asked the Kennel Club to authorize this point of view.[91] The issue quickly caught the attention of dog people in Britain and North America, and it became inflamed because a few crossbred Collies had somehow managed to get registered in the English Kennel Club. Mason believed these wrongly registered dogs belonged to the Shetland breed. Purity remained an issue, and foreign blood might contaminate the Collie through the Shetland dog, which had descended from spaniels.[92]

Others, however, favored the little dogs' maintaining the name Collie. "Personally, I cannot see why the bringing of this Toy forward should in any way trouble ordinary Collie fanciers," one person wrote to Mason. "You, perhaps the foremost authority of our present Collie, will, I am certain, admit that the Collie of today is a vastly improved dog of a more graceful appearance than his forefathers, which has only been the result of skilful mating and breeding—then, why shouldn't this little dog have the opportunity of making himself presentable." Another writer said, "It is with surprise I learn that Mr. W. E. Mason is now promoting opposition to Shetland Collies, and was responsible for a motion to report them to the Kennel Club for daring to possess their only name. One begins to wonder who and what next Mr. Mason will attempt to crush? Surely the Kennel Club will not respond to this move to crush the aspirations of a perfectly legitimate fancy."[93] Some noted that no objection to the use of the Collie name had arisen in Scotland, where all the Shetland shows had been and where the Collie had in fact originated. Only the English specialty clubs complained.

Another writer to the *Collie Folio* pointed out that Collie fanciers did not intend to crush a variety, and "that Collie men [were not] jealous of the advancing popularity of this rival to the supremacy of the Collie." It was simply that the Shetland had not been bred from Collies and bore no resemblance to that breed. "All that we claim is that the name of the Collie must not be perjured."[94] In the meantime a Shetland Collie Club was formed and applied for Kennel Club recognition as a specialty club. The person who represented the Shetland Islands in Parliament took a keen interest in the controversy.[95] By then it had become clear that the wrongly registered Collies had not been Shetlands and that the registration was a mistake, not a matter of deceit. Even so, in the spring of 1909 the Kennel Club had decided to stop recording Shetland Collies, on the basis that the dogs did not look like Collies and had not descended from Collies. A reaction followed immediately. The Ladies' Kennel Association show planned to have four classes for Shetlands, and breeders of these dogs needed pedigree papers to exhibit. These people began to write to the Kennel Club, asking what position the club took on the status of the Shetland as a breed and also asking whether they could show the dogs. The Shetland fancy would be hamstrung by the Kennel Club's action. To add to the difficulty, the judge of the Ladies' Association show was the wife of the M.P. for the Shetlands, and she had bred Shetland dogs all her life.[96] The Kennel Club, however, decided that the name Collie could not be used for the Shetland breed, nor was the specialty club founded for the breed allowed to use the name. Mason fervently wanted the Kennel Club to stand by its position. "It is devoutly to be hoped," he wrote, "that the Kennel Club, who are in the best position to adjudicate on the matter, will adhere to the stand they have so wisely taken, which, we may remind our readers, is that for the present none of these dogs shall be registered as Collies or even as Shetland Collies, and that the application for the Club title be also refused."[97] In the end Collie breeders did win out, and the Shetland became the Shetland Sheepdog. Breeding of the Shetland Sheepdog, popularly known as the Sheltie, to modern standards resulted from the introduction of Collie blood, and this little dog came to be known popularly as the miniature Collie. The breed's name, however, remained Sheepdog, not Collie. Mason's campaign had had lasting effects. The whole controversy, however, illustrated obsession with purity from other breeds, but even more, the sense that the name Collie itself commanded market status.

As the show Collie developed, other contentious issues arose within the ranks of breeders themselves. Color, for example, quickly became associated with ideas about purity in the breed, and later even quality. Certain colors seemed to be evidence of foreign blood. Early dogs had been small and were normally black or merle (a mottled blue gray). Color patterns changed quickly, raising issues of purity in the minds of some breeders. Black with tan markings had become popular by 1870, and many believed that Gordon Setters had introduced the color pattern.[98] In 1877 the *Canadian Poultry Review* carried an article called "The Colley," basically taken from the writing of a British Collie expert:

> Some years ago constant descriptions of colleys, with beautiful jet black coats and rich orange tan markings were given, and in advertisements and elsewhere we still occasionally hear the reverberation of the silly-song. What stronger incentive could there be to dealers to offer half-bred Gordon Setters as pure colleys ... [than] a demand for the graceful mongrels, with thin coats "soft as a lady's hand," feathered legs, draggle tails, saddle-flap ears, and a rich mahogany-color kissing spot on each cheek, that have been so plentiful ever since.[99]

Other dog experts argued that the Collie influenced setter coloring.[100] Hot debates centered on the setter issue and went on for some time. Dogs in the show ring that carried the low setter ears—not uncommon for any Collie before 1890—were normally accused of having setter blood. "The setter type is to be avoided, with its pendulous ears," noted a prominent dog scholar as late as 1927.[101] To this day it is not entirely clear what role the Gordon Setter played in the makeup of the modern Collie.[102] With or without setter blood, tan in Collies became common in tricolor (black, tan markings, and white).

In spite of the conflicts that arose out of Collie breeding strategies between 1880 and 1920, Collies—and even the progeny of top show Collies—continued to attract admirers. The late nineteenth-century Canadian farm press reflects these patterns in North America. It regularly reported on the use of show Collie genetics in producing the farmer's dog, not the breeder's or fancier's. A reader from Winnipeg, Manitoba, who owned a dog descended from Morgan's Wishaw Clinker, wrote to the *Farmer's Advocate* saying that "a farm dog should be a general purpose one, and there is no breed that I am familiar with that is so well suited for the work as the collie."[103] In 1899 the *Farmer's Advocate* offered a Collie puppy, bred by a notable Ontario Collie breeder who imported good dogs—descending from such lines as Ormskirk Emerald's—to anyone sending the names of twelve new subscribers.[104] *Farming* carried a picture of two Collies in 1895 and added: "Very closely connected with improved farm stock is the handsome and intelligent collie. The beautiful pair that are depicted in the accompanying illustrations are specimens of a sort that has become [as] fashionable as a lady's companion in the drawing room as it is popular among the shepherds on the hillsides."[105]

People generally seemed to be aware of and interested in the best show dogs of the time, and the progeny of some of these dogs could even reach ordinary farmers. The special relationship between Collies and the public emerges in literature that appeared in the early twentieth century. In 1921 Katharine Lee Bates published *Sigurd, Our Golden Collie and Other Comrades of the Road,* in which she described the charming adventures of a pet Collie. One of the most striking characteristics of the book was that she believed a detailed account of her dog's breeding would interest her readers and be understandable to them because they knew about Collie affairs. She traced Sigurd's ancestry to Stracathro Ralph, one of the important sons of Christopher and a direct descendant of Charlemagne. She also mentioned the great Christopher himself, as well as his sire, Metchley Wonder.[106] Bates spoke knowledgably about desirable show Collie style when she described Sigurd. "His happy body, gleaming pure gold in the sun, with its snowy, tossing ruff, was both too tall and too long. His white-tipped tail was too luxuriantly splendid. The cock of his shining ears was not of the latest kennel style. His honest muzzle was a trifle blunt. He was, in short, lacking various fine points of collie elegance."[107]

Albert Payson Terhune (1872–1942) was an author closely associated with the love of Collies and their culture. Terhune began writing about his Collies in 1915 when he published a short story about his beloved dog Lad, born in 1902. "Lad was an eighty-pound collie, thoroughbred in spirit as well as in blood. He had the benign dignity that was a heritage from endless generations of high-strain ancestors," Terhune wrote in 1919.[108] That was only the beginning of Terhune's involvement with the breed. He became a breeder of Collies and eventually even a director of the Kennel Club. Until the late 1930s, Terhune continued to write extensively about the Collies he bred and owned. He opposed the Anfield Model temperamental type, but in the end he used it because he wanted to produce successful show Collies. His Sunnybank Kennels housed several champions, one of them descended from Weardale Lord (known in the United States

Albert Payson Terhune with three of his famous dogs. This photograph shows Terhune playing with Lad (1902–18, in the background), Bruce (1909–20, in the foreground), and Wolf (1913–23, a son of Lad's, on his hind legs). Terhune described Wolf as a throwback to wolf ancestors, showing that he shared other Collie breeders' view that the Collie related particularly closely to the wolf. This theory cannot be substantiated. The dog named Wolf did, however, look like an old-fashioned Collie. Note the head shape and the short fur coat.

as Knocklayde King Hector), a dog of uncertain temperament from the Model line. Terhune greatly admired such dogs as Squire of Tytton and Southport Sample, and he likened his great pet, Sunnybank Lad, to Sample. But as Terhune's breeding program became more sophisticated, he turned away from the Sample type. Some of the best show Collies he bred resulted from a cross of the Weardale Lord blood in Sunnybank Sigurd (named after Katharine Lee Bates's dog) on a bitch of Canadian breeding, Alton Andeen.[109]

His stories described what he disapproved of in large kennel operations. He dismissed rich breeders as men who knew little about dogs or their breeding, and he typified this type of fancier in a fictional character called the Wall Street Farmer. Terhune's character knew nothing about the quality of his dogs, and he went to great lengths to import extremely expensive Collies from Britain for the sole purpose of winning at local dog shows. One cannot help wondering if Terhune had J. P. Morgan in mind. He certainly discussed Morgan, his dogs, and their show careers in some less well-known articles.[110] Real dog experts, he believed, spent their lives perfecting the challenging art of breeding Collies. Terhune also commented on savagery in expensive dogs. He wrote ambiguously, however, both about savagery and about the intelligence of show Collies. He suggested that valuable dogs in the large kennels behaved stupidly and even acted mean largely because of poor training in such an environment. Since Terhune attributed higher intelligence to Collies bred from important show stock, and because he himself bred from show lines that carried savagery, he tended to underplay stupid and vicious behavior in his novels. He was trying to attach the mental attributes he admired in the Sample type to the show style that he increasingly bred from by the late 1920s. He therefore took no clear stand on the issue of breeding for utility or temperament versus fancy or beauty and on how such breeding programs affected quality. For example, Terhune attached specific characteristics denoting excellence to Collies of good breeding and claimed that only purity of breeding could produce them. For him, beauty, agility, brains, courage, loyalty, and honor were all qualities instilled in Collies through purebred breeding for fancy points. At the same time, however, Terhune railed against the effects of wealth and shows on the quality of the dogs so bred.

Even so, Terhune understood how intimately markets and the ability to sell purebred animals related to the desire to produce them. In an article named "The Collie: There Is Money in Him," Terhune elaborated on the role markets play in breeding: "The amateur rural-sojourner can make his collies pay the home's taxes and bring in a little surplus, besides. I know, because I have tried it." "There is a joy in [breeding], too; and sometimes an excitement worthy of the race track."[111] He made some interesting observations about the American dog breeding world in the 1920s:

For pedigreed dogs have ceased to be pets. They are the basis of a recognized and daily bigger business; a business whereby thousands of people are making a living. . . . To every American who has reduced the dog business to a profitable science, there are probably five Europeans who are doing the same thing. There are more women in this business that in any other line of financial endeavor of the same volume. . . . There is no comparison between the sales of dogs in Great Britain and in America. The Briton long ago established a worldwide market for purebreds; a market with which we are just beginning to compete. England and Scotland still have the inside track on this. . . . The continent of Europe gives Great Britain just as good an export market in dogs as we do. Indeed, Continental buyers often outbid us Yankees for some exceptionally fine show-specimens.[112]

In fact a great deal of the culture behind the motivation to produce purebred animals emerged not just in nonfiction passages like those above, but in his stories as well.

Terhune believed the Collie was the oldest of dog types, and he repeatedly claimed that the breed related closely to the wolf. He concerned himself with genealogy, and he regularly discussed the ancestry of the dogs he wrote about. In his dog stories he linked purity, purebred breeding, and general excellence of character, and for Terhune all these characteristics could be associated with the best and purest aspects of humanity. The subtle, and often not so subtle, connection between dog values and human values in his books suggested a commonality in issues of dog breeding and human breeding. Terhune's books breathed Collie culture, and he discussed the history of many dogs described in the *Collie Folio.* His love of Collies inspired many people to breed them.

Collie culture in his books often had a deep and lasting influence on readers in a way not necessarily related to dogs. Terhune's biographer, Irving Litvag, is an example. He wrote about his pilgrimage to the ancient house of Sunnybank, long after Terhune's death and just before the dwelling was torn down: "After so many years of dreaming about it, he finally had come home—home to his childhood, home to the friends of his boyhood dreams. Home to the sounds of barking collies, to heroic deeds, to twilight talks on the veranda overlooking quiet water. Home to a world he had longed to find and never could. He had come home to Sunnybank. And he had never been here before."[113] As Litvag wandered over the Sunnybank property that cold December day in the 1960s, he suddenly came across Lad's grave, which Terhune had eloquently described in his writings. Litvag described how he "looked down again at Lad's grave and suddenly found himself crying. . . . Why in the world would the grave of an animal have this effect on him? The tears abruptly ceased as he wondered about it. Was he crying for his lost childhood? Was he crying for his dead parents? Was he mourning for old heroes, long-lapsed fable, ancient dreams that disappeared as dreams always do?"[114] Litvag's deep emotions about the Terhune Collies resulted from sentiment rooted in something more than a love of dog stories. Collies carried attributes that personified human values. Terhune's work inspired a vision that might be described as fantasy or even myth. Litvag's words show how Collies could become a symbol for social values, and social values probably lay behind some people's motivation to breed the dogs.

The huge success of Eric Knight's *Lassie Come Home* (published in 1940 and based on a short story that appeared in 1938) demonstrated the enduring popularity of the Collie that existed by the mid-twentieth century. But the public's special excitement about the breed's genealogy had vanished. The extension of the Lassie cult beyond the book, perpetuated in movies, television, and comic books, clearly showed that the breed remained popular, but the concern with Collie breeding culture evident in the first part of the century and in earlier literary works was not part of Lassie's appeal. Concern with Collie breeding, if not Collie dogs, would increasingly be restricted to those who bred them.

Collie breeding patterns reveal problems embedded in any breeding system based on pedigrees. While foundation dogs often came from an unknown background, for example, devotion to pedigree and the idea that it preserved purity and quality persisted. Exactly what did a pedigree mean or guarantee? How did it reflect an animal's heredity? Problems over the use of foreign blood and over the right to use the Collie name arose out of these difficulties. Collie breeding also demonstrated that breeding purely for fancy points made defining improvement more difficult, because fundamentals relating to original or authentic type were threatened by open disregard for utility. The deep-seated perception that breeding for fancy undermined the general utility of the dogs created a dichotomy: Could show Collies be described as "improved"? Many believed the new Collie was a degenerate version of the original type. Elitism and money also seemed to play a part in that degeneration. Color could influence breeding decisions, and some believed color reflected underlying characteristics such as overall show quality, trueness to original type, or freedom

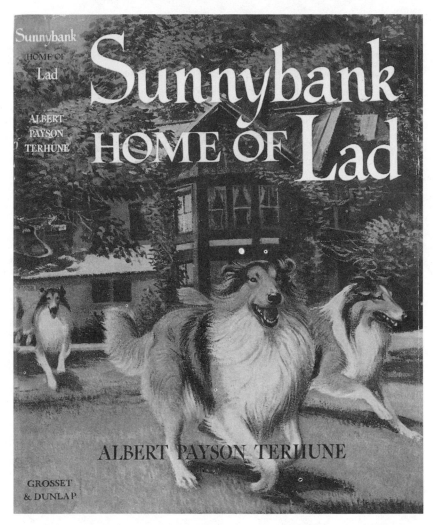

A popular image of Sunnybank Collies that adorned the book jacket of the Grosset and Dunlap edition of *Sunnybank: Home of Lad* in the 1960s. In his biography of Terhune, Irving Litvag spoke of this picture, with its glowing red gold Collies running excitedly over the emerald lawn. The image came poignantly to his mind when he visited the deserted Sunnybank property.

from foreign blood. In spite of these conflicts—perhaps surprisingly—the Collie remained a popular breed.

Because this book deals with a breeding system that originated in Britain, logic suggests that the method would be practiced on animals native to that country and that I should concentrate on the development of British breeds. But in order to see how ubiquitous some aspects of the purebred breeding system could be, it is imperative to look at another animal. To help us understand just how important Britain and the United States were to world trade in purebred animals, we need to consider the history of a breed that initially developed outside those countries, was not created by the breeding system I am discussing, and had an established international trade outside centers in either country. The story of historical breeding of the Arabian horse therefore follows.

# Five

# A WORLD MARKET FOR ARABIANS
# TAKES SHAPE

"The exploits of the Arabian have been sung by Arabian poets, and handed down from generation to generation over the midnight camp fires.... [The Arabian] is the chief and noblest origin of our national racehorse, and of the best breeds of North Africa and of light breeds all over the world," wrote Lady Wentworth. "It is essential that we should ... preserve a foundation stock of perfectly pure Arabian blood of the highest type and strains, rejecting vigorously all doubtful blood."[1] In these words Lady Wentworth (Judith Blunt Lytton), one of the most significant European breeders of Arabians in the twentieth century, captured the Western vision of the horse—strength of type as a result of purity maintained by Eastern breeding. Lady Wentworth's comments also introduce us to some of the basic conflicts that developed in the Western breeding of Arabians. From earliest times in Europe, the horses filled both a utility role and a fancy role. Arabians helped to create racehorses and also came to be valued for infusing endurance into other horses—an essential attribute for cavalry mounts. But the wish to maintain the purity of their original Eastern breeding could itself be described as breeding for fancy: perpetuation for preservation, if not for beauty points. Fancy in this case would bring out deeper problems with respect to authenticity of type and how to define what made an Arabian pure.

In 1958 an American expert raised some of these points: "Dazzled by the fiery elegance, the stylish bearing and the fabled history of the Arabian horse, some fanciers [note his use of this word] of the breed have naively insisted that the Arab is and always has been 'perfect' and that any attempt to improve on that perfection would be presumptuous."[2] All Arabian horse breeding outside Arabia would ultimately demonstrate difficulties that emerged from a fusion of Eastern and Western thinking. Western and Eastern breeding culture would become linked in a particularly significant way when a trade in "purebred" Arabians developed between Britain and the United States. That connection ultimately led to a global

Arabian market. The breed's global impact makes it particularly interesting to study in relation to many of the issues addressed in this book: the role of trade in purebred breeding, the effects of the market on perceptions of quality and purity, the motivation to breed for utility or fancy, the relation of authenticity to "improvement," and the role of public herd books in regulating all of these factors.

Attempts to understand the origins of the Arabian horse have triggered many legends—"some as charming as fairy tales," as one author noted—to explain the horse's beginnings. The breed existed at least as early as 1635 B.C., and some believe the Arab people created the horse on the grassy slopes of the Caucasus Mountains before the animals existed in the desert.[3] Others have suggested that the Arabian horse could be found in Egypt as early as 2000 B.C.[4] Scholars have also argued that the hot climate played a role in fixing type, and type of a specific sort, in the Arabian.[5] Wherever, and however, the "breed" was fixed, we know that most Arabian horses after that time were bred by Bedouin tribes in North Africa, in an area bordered by Nejd in the south and the Syrian desert in the north.[6] All agree that Nejd itself was the most significant center, but there is some difficulty in locating the exact geographic position of historical "Nejd," today sometimes called "Najd." Some Arabian horse breeders place it in parts of Syria, Iran, and Iraq.[7] The diaries of Lady Anne Blunt from a trip to Arabia in 1879, taken in their entirety, suggest that while Bedouins produced stock in those areas, the greatest historical horse breeding tribes of Nejd were in present-day Saudi Arabia.

Arabs' breeding techniques have for centuries fascinated Turks, Egyptians, western Europeans, and Americans.[8] In spite of considerable knowledge about the culture surrounding them, Western breeders have not held (and do not hold) common views on what those breeding methods were. We do know that after a "breed" had been established (by climate or selective inbreeding or both), the Bedouin breeders kept track of what were called "strains" in the horses. Strains defined the family line an individual horse belonged to, revealing its genealogy in a certain way. Three overarching strains (all with numerous substrains) came to represent three distinct physical types within the single breed. The three strain-types were the Saklawi (refined, elegant, and feminine), the Kuhaylan (powerful, bold, and masculine), and the Muniqi (with a racier build and unusually with the forequarters more developed than the hindquarters).[9] (These names can all be spelled in a variety of ways.) One particularly significant characteristic of Arab strain recording was its emphasis on female lines. The

Arabs kept track of their breeding operations within the strain relation of mares, not stallions.

While we are aware that the Arabs emphasized their mares and kept records of mare families, described as strains that descended in tail female, we do not know how either mare emphasis or strain records influenced their breeding decisions. It remains unclear whether record keeping in the form of strains worked as a breeding tool. In other words, we do not know whether the Bedouins practiced strain breeding, nor do we know whether strain breeding involved either inbreeding or pure-in-strain breeding. These questions have remained critical to breeders of Arabian horses, though, from the time the animals left their desert home. The idea of strain breeding, or what might be called strain theory, became connected with ideas about purity and interacted with Western ideas about public pedigrees. The problems of defining quality, purity, and authenticity, which we have traced in cattle and dog breeding, would be much more complex with respect to the Arabian horse because ideas about strain theory would remain part of Arabian breeding culture. Most breeders developed their own version of what strain theory meant and how it should affect breeding programs.

Egypt was the first important country to import the horse of Nejd. The whole Arabian Peninsula, Iraq, Syria, Lebanon, and all of North Africa came under Turkish influence after the thirteenth century, and Egypt developed into the center of a Turkish province. The sultans of Turkey encouraged the founding of private Arabian studs in all their provinces, so that cavalry horses could be made available to the state when the need arose. As early as the fourteenth century, princes in Egypt established their own stables for breeding Arabian horses. Over the centuries many studs were founded and foundered, for Egypt was not naturally a good horse breeding country.[10] It was the nineteenth-century rulers of Egypt who created studs that endured and ultimately dispersed bloodstock to both Europe and North America. In 1819 Mohammed Ali, ruler of Egypt, responded to the Turkish sultan's orders to quell an uprising that had originated with Arab tribes in Nejd. Under the leadership of Mohammed Ibn Saoud and fired by the religious movement to purify Islam (Wahhabism), these Arabs had managed to take control of all of Nejd and other parts of central Arabia and—most important to the sultan— the holy cities of Mecca and Medina.[11] Mohammed Ali sent an army to Arabia under his son Ibrahim Pasha, who would return to Cairo victorious, bringing a number of the finest Arabian horses bred in Nejd. Mohammed Ali had always been a lover of good horses, and these captured animals formed the nucleus of a magnificent stud, housed in ornate stables, that would expand through further raids on Nejd Bedouins.

In 1823 Abbas Pasha, grandson of Mohammed Ali, became viceroy of Egypt at age twenty three. Abbas Pasha had always loved Arabians, and he had been put in charge of Egyptian horse breeding stations by his grandfather. The new viceroy threw his energy into maintaining and enlarging a stud that at its height housed one thousand horses.[12] Abbas attempted to verify the purity of his animals by sending scribes to the desert to write down pedigree information previously preserved by word of mouth. In many cases the only record that could be had was the horse's strain, or the name of the dam's family. Sires were often unknown. But he managed to set up a personal, and therefore private, studbook. Abbas also brought Bedouins back to care for his Arabians and to supervise breeding. It has been argued that Abbas Pasha altered the original Bedouin horse even while he claimed he was trying to preserve both purity and authenticity. He collected desert horses for their beauty and their ability to sprint, and he bred for those characteristics. Bedouin horses tended to reflect selection for endurance rather than short speed and for the ability to go far on little food and water.[13]

The history of these important Egyptian studs reveals that while horses outside Arabia might serve as war horses, ideas about preserving original type played an even more significant role in breeding programs. The fundamental problems of how purity related to the authenticity of original sorts found in Arabia and of what improvement for utility use meant within that framework would be embedded in Arabian horse breeding from this time on. While Abbas Pasha later received credit for preserving the authenticity of the Arabian, it seems evident that his breeding practices reflected his personal vision of how that authenticity related to quality as much as actual authenticity. Abbas bred, it appears, for what he believed was improvement of original type.

In 1854 Abbas Pasha was assassinated. His son, El Hami Pasha, took little interest in the horses and subsequently auctioned off the stud in 1860. Because breeding in Arabia had declined (owing to tribal wars, animal disease, and loss of interest in horse production), buyers from various countries who wanted Arabian horse genetics flocked to this Egyptian stud's dispersal sale. Although at least two hundred animals went to Europe, the horses left no mark on subsequent European breeding programs. Ali Pasha Sherif (whose father had been educated by Mohammed Ali and had served him as governor of Arabia) bought some forty horses, and in the end he preserved what would be the greatest and most influential part of the stud. A man of great wealth, he built elaborate stables near his palace in Cairo to house his treasured horses.[14] While ultimately he saved the bloodlines of Abbas Pasha for future Arabian horse breeders, he also preserved Abbas Pasha's vision, perhaps more than authentic type. By 1873 Ali Pasha Sherif was stabling four hundred horses that descended from the breeding of Abbas Pasha. The operation had begun to decline by 1880, however, owing to a virulent horse sickness. By ten years later the stud was devastated. Ali Pasha's sons believed the stud had eaten into their inheritance, so they imprisoned their father, who died shortly after. In 1897 the sons dispersed the remnants of the breeding center.[15] Wilfrid and Anne Blunt, important future Arabian breeders, saw the stud in its final days and had this to say about its condition: "We saw Ali Pasha's Stud in the last years of its disruption. Decimated by plague and weakened by years of inbreeding and gross neglect, the horses were of an ethereal quality and truly like gazelles, with no more bone. It was 'Type' etherealized almost to extinction."[16] Ali Pasha Sherif had apparently hoped to preserve both purity and authenticity by breeding for beauty.

The earliest important center of Arabian breeding in Europe took root in Poland. Although the Poles had begun upgrading local horses with Arabian genetics as early as 1570, it was not until the seventeenth century that the practice became widespread. Wars with Turkey in that century provided the Polish people with a supply of captured Turkish cavalry mounts of Arabian breeding. When these wars ended at the end of the century and the supply of raided horses dried up, Polish breeders began to rely on horse dealers in the Ottoman Empire or on special envoys sent to Asia Minor in search of Arabian stallions. The Poles bred these stallions to local mares, and after constant crossing back on Arabian stallions over generations, they became bred up to have a greater and greater concentration of Arabian blood.[17] A number of princely estates maintained large Arabian horse studs where the animals were bred to supply mounts to the Polish nobility. Several are important to this story. In 1791 the Sandgusko family, wealthy landowners, established the Slawuta Stud, consisting of some four hundred horses. Several generations later, a brother and sister divided the operation. The sister, Maria Sandgusko, married into the Potocki family, owners of a breeding farm at Satanov. The fortunes and genetics of these studs became intertwined. Many of the Slawuta animals would become part of the Potocki holdings. In 1881 Joseph Potocki, Maria's son, founded Antoniny near the original Slawuta stud with the existing Arabians at that time—about one hundred horses.[18] The affairs of the two family studs—Slawata and Antoniny—remained closely linked. Private breeding records existed for all of these interconnected family operations. Record keeping for Arabians in Poland at the end of the nineteenth century resembled patterns in eighteenth-century Britain, where the aristocracy kept private records of their breeding of horses and dogs. By the late nineteenth century most animal breeds in Britain had a public record book, and breeders tended to record their animals in those books, often for market-related reasons.

The imperial family of Russia and the Russian aristocracy also bred Arabians for service to the army. Arabians seem to have been imported from as early as the time of Ivan the Terrible in 1533, but no records survive from this early period. In 1778 Count Alexis Orlov founded a stud based on gifts from the sultan of Turkey.[19] More studs developed after that time, and stud records for these operations were privately kept. Trade in horses did occur between Polish studs (usually in Russian Poland) and those in Russia. Thus two significant Arabian breeding centers outside Egypt had developed in Europe by the mid-nineteenth century, of which Poland remained the more significant.

In the seventeenth and eighteenth centuries Arabians found favor in European countries other than Britain, because the need for cavalry remained greater on the Continent than in an island nation. Stallions entered Britain from 1616 onward and helped create the Thoroughbred, but Arabians in their own right did not achieve great popularity in that country. A single stud established in the 1870s, however, would make Britain a center of Arabian horse breeding from an international, if not a British, perspective. The Crabbet Arabian Stud, founded by Wilfrid and Lady Anne Blunt in 1877, would also take the world's trade in Arabian horses to new heights. Interest in the Thoroughbred initially triggered the Blunts' concern with the Arabian. Lady Anne's family bred Thoroughbreds and believed that the Darley Arabian, a stallion used in the early eighteenth century to create the English racehorse, had been of great significance in the excellence of the Thoroughbred. The desire to find another horse like the Darley Arabian, or at least stallions of the same strain, inspired the venture, and Lady Anne's brother, Lord Wentworth, agreed to help fund the Blunts' trip to Arabia. The Blunts' early interest in the Arabian, then, followed from a centuries-old pursuit in Britain.

The Blunts quickly abandoned the idea of improving Thoroughbreds through more Arabian blood. They decided instead to establish a breeding operation based on the best Arabians they could find, in order to preserve the breed.[20] The problems of purity and its relation to original type were part of their breeding strategy from the beginning. Concern for the purity of the breed, however, meant that they intended to import mares for their breeding operation and not rely on the progeny of local non-Arabian females bred to imported stallions as foundation stock. The Blunts traveled to Aleppo and met with an agent, and there they bought their first horse, Dajania, a filly bred in the desert, for £35 (the sterling equivalent of U.S. $175). Other mares followed, and a few stallions as well. Two of these early purchases had lasting effects on the breeding of Arabians: the mare Rodania, bought in the desert in 1881 for £124, and the stallion Pharaoh, purchased from Arab breeders in 1878 for £275. Rodania's blood would ultimately take Crabbet breeding all over the world, as sons of her daughters were exported from Britain. The second horse, Pharaoh, serves as an example of how the early international Arabian trade worked. The stallion, after being used at stud at Crabbet, went to Count Potocki of Poland in 1882 for £550. Potocki then sold the horse in 1885 to the Russian imperial stud, where the animal was head sire for a number of years. Like Rodania's, Pharaoh's genetics would become entrenched in the major breeding centers of the world.

The Blunts later traveled through Nejd and purchased mares celebrated by the Bedouins. The English couple paid about £100 for each of these horses. One mare, Queen of Sheba, cost £265, but they considered her special. Prices generally did not go up in their later purchases, though. The Blunts learned Arabic and collected information on strains and the history of Bedouin breeding. By the late 1880s they began buying horses from the stud of Ali Pasha Sherif. One supremely important stallion they purchased there in 1889 was Mesaoud. This stallion, probably the most significant horse they bought, cost them only £60. Mesaoud ended up in Russia; Shahwan, a gray stallion also bought at the stud, was later sold to a breeder in the United States. The Blunts attended the 1897 dispersal sale, purchased more Ali Pasha Sherif stock, and established an operation in Egypt, called Sheykh Obeyd, to breed stock and also to house Egyptian horses that later would go to Crabbet in England. From this nucleus of desert and Ali Pasha Sherif horses, acquired between 1877 and 1897, the subsequent Crabbet Arabians were bred.[21]

Purity, in the form of authentic documentation from breeders' private records, remained as important to the Blunts as good quality. In 1898, for example, they sold many horses at Sheykh Obeyd because of faulty records. The couple quickly discovered that many Egyptian dealers invented pedigrees in order to sell horses to wealthy Europeans. The Blunts went to great lengths to ensure that breeders' private documentation on all the stock was as correct as possible. They understood Bedouin breeding methods better than most Europeans, and Lady Anne left an incomplete manuscript on strains and breeding theory. The Crabbet Arabian Stud intended, however, to preserve the Eastern "purity" of the Arabian horse through documentation under British traditions. Of supreme importance, then, was a system for public registry. The General Stud Book (GSB) had always registered Arabians for reasons relating to racing.[22] In 1877 James Weatherby agreed to open a special section in the GSB to record Arabians and their pedigrees.[23] The Blunts intended to have publicly registered pedigrees that carried Eastern cultural aspects. Eastern and Western recording were combined at Crabbet in a complex way. While the horses had registered pedigrees and stallions might be emphasized in breeding operations, mare breeding lines received recognition through a naming pattern. Any horse, mare or stallion, with a name starting with R, for example, descended on its dam's side from Rodania.[24] Any horse that descended from Dajania on the dam's side had a name that started with N, for Nefisa, Dajania'a daughter. The Blunts did not strain breed their Arabians and did not use inbreeding at the stud. Believing their horses to be of fixed type and true to that fixed type, the couple bred to perpetuate horses defined as "pure," or *asil*, meaning they traced authentically on all sides to Bedouin breeding. Through the Blunts, the Arabian came into the British purebred breeding world without entirely abandoning Eastern breeding culture. In 1880 the stud produced its first foal crop, and the Blunts recorded four foals in the GSB.

From the beginning the couple relied on foreign buyers to defray the costs of running the stud, because Arabians continued to be attract little attention in England, where the Thoroughbred remained so popular. Entry in a public studbook improved their ability to sell on the international market. Tattersall held the first of many Crabbet auctions in 1882, and the Blunts asked more than two hundred people to luncheon in order to make it a social event. Some fourteen horses went into the sale, and eleven sold. Over the years, auctions at Crabbet would generate only enough to cover the cost of running the sales—usually about £500. The stud continued to be a drain on the Blunts, in spite of sales to Poland and later to Russia. It also grew in size over the years, increasing the drain. By 1892 Crabbet stabled sixty-one horses. In the early years of the stud's operations, Wilfrid Blunt tried to promote Arabians as racehorses, in the hope of creating another market. In 1884 he persuaded the Jockey Club to hold a race at Newmarket under its rules, but for Arabians only. The Jockey Club considered the race a fiasco and refused to hold more races for Arabians, so that market never materialized.[25] In spite of these difficulties, the English Arabian stud would ultimately find a good market in a new location—the United States.

Arabian horses had come into that country in small numbers at the end of the nineteenth century and early in the twentieth century, from England and the East (both Arabia and Egypt). The first Arabians to arrive in the United States were stallions presented to President Grant by the Turkish sultan in 1878. These animals impressed Randolph Huntington, the first man to breed an Arabian in the United States, and in 1888 he imported horses from Crabbet. Shortly after, men like Spencer Borden began bringing in stock from Crabbet to serve as a breeding pool for the improvement of military horses. Borden, a Harvard graduate and businessman, imported stock between 1898 and 1911. He brought in a particularly important mare, Rose of Sharon, daughter of Rodania.[26] Americans paid more attention to the breed because of its known ability to influence stamina. Arabians, they believed, could help produce good horses for the cavalry. The breed would play a significant role in remount stations, government-funded centers where horses were produced for cavalry use, and that created a market for the breed.

Interest in government remount stations that actually bred stock emerged some time before such centers materialized. Although Congress turned down Theodore Roosevelt's plan for government horse breeding operations under what was called the United States Cavalry Stud, many people still believed the Arabian could serve as a breeding tool for the army. In 1906 Homer Davenport, who wrote about breeding Collies and the effects of the dog fancy on basic quality, decided to import Arabians with the cavalry in mind. He went to the desert to seek Arabians of the greatest authenticity to type. President Roosevelt, who believed Arabians were useful for breeding army horses, provided Davenport with authoritative support, and a wealthy businessman, Peter B. Bradley, funded the endeavor. Davenport would ultimately be particularly important to American Arabian breeding for three reasons. He was concerned with authenticity from the desert horses and thus with the idea that divergence from original type meant degeneration. The subsequent breeding of the horses he brought in without other Arabian-line infusion until the herd's dispersal in 1922 (Davenport himself died in 1912) allowed other breeders to preserve "authentic type." And he played an important role in American registry of Arabians.[27] In 1906 Arabians did not have a separate registry system in the United States but, as in Britain, could hold pedigrees through the racehorse industry.

American horse racing, and the structure to support it, had evolved along British lines after the late eighteenth century, but public registration of Thoroughbreds within that system remained complicated in the United States. The American Revolution triggered the establishment of the first public American Thoroughbred record books. Various public studbooks existed in Virginia as early as the 1820s. But Thoroughbred breeders in New York and Maryland continued for some time to rely on the General Stud Book in Britain for recording. The American Jockey Club, founded in 1854 in New York, set up a new public record book for racing Thoroughbreds in the United States, but many breeders continued to use the GSB as well. (The Canadian racing system was an offshoot of the American one, with Canadian breeders relying on stock from Kentucky and Virginia.)[28] By the late nineteenth century Arabians born in the United States could be pedigreed by the American Jockey Club. Imported Arabians, like imported racehorses, could be registered if they held pedigrees in the English GSB. Arabian registry affairs, then, remained intimately linked to the Thoroughbred industry in both Britain and the United States.

Davenport failed to take his twenty-seven imported horses through Britain and acquire GSB papers. He aggravated the problem by not getting certificates from the Bedouin breeders that gave the horses' ancestry. As a result, they could not obtain pedigrees in the American Jockey Club's book. In 1906 the Bureau of Animal Industry, which decided what qualified as a purebred animal as a result of the United States Department of Agriculture's role in tariff regulations set out by the Treasury Department, compelled registration in American books for purebred status. Davenport's Arabians were not eligible for that status, because the only American book that could register Arabians would not accept them. Davenport thus found he had to look for another way to give his imports pedigrees. Otherwise, regardless of the perceived purity of their breeding by Eastern standards, they would not be recognized as pure under Western principles and would lose their monetary value. Davenport and a few supporters set out to establish a new public registry for Arabian horses. These men founded in 1908 what would ultimately (not until 1969) be called the Arabian Horse Registry of America.[29] Not all American breeders wanted a division between Thoroughbreds and Arabians with respect to recording. Borden, for example, refused to register his horses in the new book for several years.[30] (The American Jockey Club continued to pedigree Arabians until well into the twentieth century. American Arabians often held two pedigrees, and as late as 1960 some horses were double registered.)[31] The new organization established its own set of rules for recording eligibility. No breeding-up would be recognized for purebred status: all horses entered had to be proved to be descended entirely from horses bred in the desert (Davenport's Arabians would, in effect, be grandfathered in). Any future imported horses would be accepted as purebred only if their purity could be authenticated. The mission of the Western registry system was thus to guarantee the Eastern purity of the breed.[32]

The international market for Arabians escalated appreciably after the United States Remount set up breeding centers in 1918. The horrors of the First World War, including the enormous loss of horses, evidently affected the thinking of Congress.[33] Growing American demand for the breed to produce remounts coincided with expansion and changes taking place at the Crabbet Arabian Stud. The operation passed to the hands of Wilfrid and Anne Blunt's daughter, Judith Blunt Lytton, who inherited the title of Lady Wentworth from her mother. She went on to become, between 1920 and 1950, one of the most renowned breeders of Arabians in the world. She found she could sell her horses on the strength of her breeding programs, based on faith in their purity and their quality. Since showing of Arabian horses was not extensive in Britain at this time, to develop the market for Arabians she was forced to rely on her promotional work, the generally perceived excellence of her stock, and the use of a public registry. She followed her parents' basic approach, breeding for Eastern purity by maintaining lines but not inbreeding. While she kept track of her female families, she emphasized sires in breeding. She considered pedigrees in the GSB to be essential for both breeding and marketing. Lady Wentworth did not breed by strain, and she believed that the Bedouins had not done so either. She argued that strain names had originally been useful, though, because they showed the location and name of the breeder (certain areas had a reputation for celebrated Arabians).[34]

Lady Wentworth attempted to alter original type, even if within pure lines, because she bred for improvement. When some accused her of diverging from authentic style, she responded: "If being 'off-type' means having higher withers, longer rein and curve of neck, stronger and broader quarters and better hocks, together with freer hock action, then I am proud to say that I have deliberately bred for these points and shall continue to do so."[35] She based her breeding program on Blunt-bred mares (from the Arabian desert and Egypt) and one particular stallion, a Polish horse named Skowronek. The gray stallion, foaled in 1908, came into her possession in 1920. Skowronek did not serve mares besides those at Crabbet, and he sired fewer than fifty foals in his lifetime. Lady Wentworth

Lady Wentworth and Skowronek at the Richmond Show, England, 1924. Lady Wentworth became one of the world's most renowned breeders of Arabian horses. Her importance to the breed cannot be overestimated, nor can that of Skowronek, the horse she made so famous. This Polish-bred stallion's bloodlines can be found all over the world, through his breeding work in England at the Crabbet Arabian Stud.

used the stallion on the old R and N lines at Crabbet to produce horses that she claimed represented the utmost purity—meaning they traced to the best of desert breeding—and also improved quality.

Three Skowronek sons in particular are important to this story. Naseem, a stallion from the N line of Dajania, went to Russia. Naseem's presence after 1936 at the government stud, Tersk (founded in 1921, with a public registry, on the estate of Count Stroganov and initiated by the Russian government to produce horses for the army), demonstrated that Russia figured in the international breeding world and also ensured that both Polish and Crabbet lines would permeate Russian breeding. Raffles and Raseyn, both from the R line of Rodania, went to the United States.

After his importation in 1932 Raffles spent most of his breeding career at Roger Selby's farm in Ohio, where a manager ran the breeding operations. Selby kept Arabians simply because he liked them, and he found Raffles particularly beautiful. The small gray stallion, just over thirteen hands and therefore pony size, was the only truly inbred horse to be produced at Crabbet—son of Skowronek and a Skowronek daughter, Rifala. Raffles was not the only horse that Selby imported from Crabbet. Between 1928 and 1933, Selby brought seven stallions and thirteen mares from that stud to the United States. Much of the blood Selby imported came from Skowronek progeny. Daughters of Skowronek subsequently went to Raffles. Extensive inbreeding to Raffles himself intensified this Skowronek inbreeding. (Raffles sired 123 foals in his lifetime.) By the 1950s, even after his death in 1953, many American Arabians would be heavily inbred through his descendants. Breeders attempted to keep the progeny as close as possible in appearance to Skowronek himself, or to Raffles.

A similar breeding approach fell to Raseyn. This gray stallion served in California after being imported in 1926 by Will K. Kellogg of Michigan, known for his production of cornflakes. In 1925 Kellogg had bought his dream ranch near Pomona, California. As a boy, he had owned a pony that he believed had some Arabian blood. He had long wanted to breed horses, and his boyhood memory made him interested in Arabians, but he knew little about the breed. He turned to Herbert H. Reese for advice and then hired him as a manager. In effect, this man became the breeder at the Kellogg ranch, not Kellogg himself. Reese ran the ranch's Arabian horse breeding from 1927 to 1939. Until 1932 he was employed by Kellogg and did the buying of horses for him. In that year Kellogg donated the ranch and its stock of eighty-seven Arabians to the University of California, on condition that the Arabian horse breeding program would continue. The operation became known as the W. K. Kellogg Institute of Animal Husbandry. In 1943 the Remount Service took over the stud and ran it until 1948. The United States Department of Agriculture acquired the stud in 1949 and returned it to the W. K. Kellogg Foundation. Later that year the foundation gave the property and horses to the state of California. The ranch became part of California State Polytechnic College, which was made a state university in 1972. The institution houses an important library devoted to the Arabian horse.

Kellogg's involvement with Arabian horses and Raseyn's story not only reflect Skowronek's impact on American breeding in these years but also show how much Lady Wentworth herself influenced the thinking of American breeders. Before 1927 Reese had worked for the U.S. Remount Service and so had some knowledge of Arabians. He learned a great deal more in the subsequent years from Lady Wentworth, and though Reese experimented with different lines of Arabians, in the end all the horses at the Kellogg ranch traced to Crabbet roots. (The ranch imported six stallions and fourteen mares from Crabbet.) Heavily influenced by Lady

Wentworth's thinking, Reese wrote extensively about the need to improve Arabians, and therefore to change them: "It is indeed fortunate for the Arab that a few of the larger breeders recognized the fact that certain faults were common in the breed, and along with striving to add further 'beauty points,' these breeders finally are able to eliminate conformation deficiencies from at least the top specimens. . . . Arabians have been bred for thousands of years, but it is only recently that so many really well-conformed, hard-to-fault, and typical specimens can be found."[36] Raseyn was one of the first horses Reese persuaded Kellogg to buy from Crabbet,[37] and he was an important sire at the ranch, producing 135 foals in his lifetime. Although not inbred himself, Raseyn was inbred to while at stud, as were his progeny after his death.

After 1950 Raseyn stock was commonly bred to Raffles stock in order to inbreed more closely to Skowronek.[38] Arabian breeders attempted to preserve the greatest intensity possible in each new generation, concentrating the blood of one particular animal through two of his sons. Inbreeding to Skowronek also demonstrated the emphasis that breeders in the United States put on male lines. It was stallion worship taken to extremes. Advertisements for Arabians that appeared in journals often did not even give an animal's sire or dam but just gave the inbreeding index to Skowronek, to Raffles, to Raseyn, or to all three. Even contemporaries noted American breeders' emphasis on the male side. It seemed to some people that American horses simply had no dams.[39]

By 1960 Skowronek had created by far the greatest sire line of Arabians in the United States.[40] At the end of the twentieth century he still appeared, although well back in the records through Raffles and Raseyn, in 90 percent of the Registry's pedigrees.[41] The ubiquity of Raseyn and Raffles blood, even if generations back by 2000, shows how completely Crabbet breeding at one time dominated American production of Arabian horses. So do other statistics from the Registry. Almost 40 percent of the horses recorded in the Arabian Horse Registry of America between 1944 and 1948 descended from stallions of 100 percent Crabbet breeding, and 70 percent of them descended from horses of over 50 percent Crabbet breeding. That situation would not change until the late 1950s, when the entry of different bloodlines in breeding programs reduced the immediate influence of Crabbet.[42] It is worth noting the monetary value of Arabians in the period 1940 to 1960, when Skowronek and Crabbet maintained hegemony over the breeding of American Arabians. In the 1940s Indraff, a very high quality stallion bred by Selby and sired by Raffles, sold for a record price of $10,000. Ferzon, a magnificent yearling stud colt also heavily inbred to Skowronek, brought $10,000 in 1953, an unusual price for so young a horse. Arabians generally did not bring anything like that amount of money. Prices did not increase much from this level before the late 1960s. Prices for good, average horses stayed at about $2,000. For example, auction sale averages from Al-Marah farm, a very large breeding operation at that time, remained near $2,600 on about fifty head offered. Champion stallions could bring $6,000. Even the best stallions rarely commanded more than $200 for stud fees in 1961.[43]

Even when Skowronek's star was at its brightest, though, change was in the wind. After American breeders learned more about other European Arabians, particularly those in Poland, during the Second World War, new genetics began to enter the United States. By 1941 the Arabian horse situation in Poland had changed profoundly from its old nineteenth-century base of private operations run by large landowners.

The First World War shattered the princely studs. New breeding centers came into existence after the war under the Polish government's minister of agriculture. Surviving horses from the old private studs served as the foundation stock for the new operations.[44] Dr. Edward Henryk Skorkowski, under the government's direction, oversaw their organization. In 1924–25 he also went to England, observed horse breeding at Crabbet, and learned how public registry systems functioned. His work led to the establishment of the Polish Arab Horse Breeding Society, founded in 1926 and modeled on the British system.[45] With new structures in place, Arabian horse breeding in Poland recovered from the effects of war. The year 1938 saw the birth of three significant stallions at the government studs— Witez II, Witraz, and Wielki Szlem—all sired by Ofir, a horse descended from an imported desert stallion and a Polish Arabian mare.[46] These three animals would have an enormous influence on world breeding of Arabians by the 1960s. Witez II did so through his living example and through his progeny; the other two did so through their progeny only.

Of the three, only Witez II, bred at the Janow stud, reached the United States. In effect, he was a war refugee. When Hitler invaded Poland in 1939, the young horse was moved with others for safety, but Witez and his handler found themselves separated from the main group by bombs. The groom hid his charge and stole food for the animal for some time. Ultimately Witez's groom surrendered the horse to the Germans, in the hope that they would recognize the stallion's worth. They did, and they sent Witez to serve at stud, as part of a plan to produce "superhorses" for Hitler's "supermen." Near the end of the war, Witez lived at Hostau in Czechoslovakia, where he was surrendered to General George Patton Jr.

Witez II, a World War II refugee. Witez, foaled in 1939 in Poland, came to the United States in 1945. Witez introduced American breeders to ancient breeding centers in Europe. Photo from the *Arab Horse Journal*, February 1961.

The Germans had decided to relinquish the stallion lest the starving Russian troops slaughter him for food. Patton appreciated the horse's value, and on 28 April 1945 Americans rode Witez and others some two hundred miles to Bavaria.

From there Witez was shipped overseas to the United States, to serve at the Remount Station at Kellogg's old California ranch. When the Remount Service was disbanded in 1949, the stallion came into the possession of Earl E. Hurlbutt. Breeders greatly admired Witez II, and his progeny did well within the rapidly developing American show structure. He died in 1966 after siring 215 foals in the United States.[47] Witez II thus reintroduced many American breeders to Polish breeding. And in the years immediately following his arrival in the United States, Arabian horse breeding in Poland rebounded from the devastating effects of the Second World War (the Janow Stud, for example, had lost 80 percent of its stock). From a base of only fifty-nine mares and eight stallions collected in 1946, the government centers restructured breeding operations.[48]

By the late 1950s importing of Arabians to the United States from countries other than Britain had begun in earnest. (Lady Wentworth's death in 1957 played a role in the decline of popularity of Crabbet stock. Cecil Covey, son of her head groom, inherited the stud and disbanded it in 1972.) Although Americans did not import British Arabians after 1960 to the extent they had earlier, Arabian breeding in Britain played a subtle role in what was developing into a global market based on American demand. British breeders often acted as intermediaries in buying Arabians from other breeding centers. The British breeder Patricia Lindsay was instrumental, for example, in many American purchases of Arabians from Poland in the early 1960s. She also imported Arabians from Poland to England and exported stock to other parts of Europe, tightening a European breeding pool. British breeders also initiated a move to a more consolidated global Arabian horse association in these years, being the first to press for a world organization. When the World Arabian Horse Organization came into existence it would be headquartered in Britain.

While some important breeders—Frank McCoy, Bazy Tankersley, and Dan Gainey being a few—continued to breed mainly within the Crabbet-Skowronek families, many Arabian breeders became excited by the "new genetics."[49] New Arabian lines were particularly appealing because of a show system that was in place in the United States by 1960. Small Arabian fairs, with classes for conformation and performance, had existed as early as the 1920s. Wealth would lead to two major hubs for Arabian horse shows. One took root in California. Extremely wealthy owners of Arabians competed with each other in private shows in Santa Barbara. Shortly after the First World War, wealthy patrons flocked to the Belvedere Hotel in Santa Barbara for winter holidays and brought their horses along to enjoy on vacation. The hotel owner built facilities for his patrons' horses, such as a racetrack facing the Pacific. Within a few years, a few residents of Santa Barbara decided to expand these horse-related activities, and they organized a larger exhibition that had classes for various breeds including Arabians.[50] From these beginnings, extensive showing had developed in California by the 1950s. A similar situation evolved in Phoenix, Arizona. Wealthy people brought their Arabians to the Biltmore Hotel (owned by a wealthy Arabian breeder), and in 1954 the first exhibition took place on the grounds of the hotel. Permanent Arabian centers for breeding came shortly afterward at nearby Scottsdale, and more elaborate showing began on private estates in that area. Scottsdale would become a major hub of the Arabian horse world.[51] Throughout this period generally, other all-Arabian shows were developing in various parts of the United States. In 1958 the first show to be designated the national United States Arabian Championships took place.[52]

In the early 1960s Eugene LaCroix, a doctor from the American Northwest who had been breeding and showing Arabians based on Raffles, liked Witez II's style, and he watched recent imports from Poland with particular interest. He wondered if these new genetics might serve the show system well, and he also remembered that Skowronek had been bred in Poland. Born on a farm in Kansas, LaCroix had been involved with horses from early childhood. During the depression, with $38 in his pocket, he set out to earn a medical degree at the University of Kansas, and soon after doing so he became particularly interested in Arabians. While practicing medicine in California he saw Arabians at the old Kellogg ranch. Later he studied more about the breed and bought Arabians, and in 1950 he began showing them. By 1956 he and his family had moved to Scottsdale, Arizona, for health reasons—one son suffered from asthma—and he quickly became part of the Arizona Arabian horse world.[53] When LaCroix decided to seek new lines in Poland for breeding and showing, he found the stallion Bask, sired by Witraz, one of the three 1938 W stallions. What happened because of that meeting changed the world of Arabian horses.

Bask had not been used at stud in Poland for a number of reasons. The state farms had restricted their output of stock owing to continuing poor postwar markets; the stallion did not command Polish admiration; and the Poles had a surplus of Witraz sons to use for breeding. Bask had been sent instead to the racetrack, the fate of many Polish horses, but after four years he had not attained a stunning career. All these factors made the horse expendable, and LaCroix managed to import Bask. The stallion went on to win national championships in the United States for both conformation and performance. His success in the American show ring validated his worth, ensuring that his progeny would be in high demand (he sired 1,046 foals in his lifetime). Bask progeny remained valuable long after his death in 1979.

LaCroix's promotional work changed the history of Arabian horse breeding all over the world, and over fifteen years the stallion Bask singlehandedly helped LaCroix build an empire, Lasma Corporation.[54] Bask's ability to produce money proved phenomenal. Some 365 of the stallion's

The immortal Bask, a Polish stallion imported to the United States in 1963 by Dr. Eugene LaCroix. The horse generated millions of dollars and played a major role in an Arabian horse boom that evolved over the 1970s and 1980s. Bask died in 1979 and became a legend. Note that the faded background emphasizes that sense. Johnny Johnston photograph.

progeny had been offered at public auction by 1987, for example, and generated over $64 million. Five had sold for over $1 million each, and Bask progeny averaged $176,000.[55] Another example of Bask's power to bring in money, under LaCroix's management, could be seen in the buying of one individual. By 1988 Thomas Chauncey had bought forty-four Bask daughters at public auction for a total of $13 million.[56] LaCroix would in effect, through Bask initially and then through other horses, establish another modern "strain," the straight Polish horse. LaCroix argued that Polish Arabians bred truly as a result of hundreds of years of breeding programs in that country, and he set out to maintain these qualities by breeding down sire lines. Mares were selected by their sires' families.[57] LaCroix marketed his Polish Arabians through auctions. He would have far greater success with that device than had the Blunts at the end of the nineteenth century.

In 1971 Lasma held its first auction, with twenty-six lots selling for an average of $20,000 apiece. The second sale, in 1974, averaged $30,000. By 1976 LaCroix began new business ventures in horse transactions. He ran auctions for other breeders. He also created a special sale for the progeny of Bask, known as the Bask Classic. Sales for yearlings and for pure Polish Arabians became part of Lasma operations as well. By 1981 prices averaged over $100,000 a head at these events. Two years later prices went even higher, and Lasma also managed more sales each year. In 1983 Lasma produced three auctions that averaged between $200,000 and $300,000 a head. The corporation also sold and bought horses for individuals under private contract. By 1983 Lasma Corporation either controlled or had interests in breeding operations in Florida, Kentucky, and Arizona. The corporation did $50 million gross in Arabian horse transactions in 1985.[58] (Chauncey, buyer of Bask daughters, claimed in 1984 that his two hundred Arabians would generate $34 million on the open market.)[59] This, however, represented only part of the empire that Bask built. The organization branched out into real estate and built an equestrian park on five thousand acres (a community of residences with facilities for horses), with land sales in that park of more than $40 million.[60] Other Lasma horse operations developed too. For example, at their estate in Kentucky, named Croixhill, LaCroix's son and daughter-in-law set up a new company offering services for new horse owners.[61]

Lasma's affairs demonstrate that a great deal of money could be spent on Arabians, and sale averages do not reflect how high prices for individual animals might be. At the Lasma V sale in 1983, one mare brought $1.5 million. Two years later she sold to a new Arabian owner for twice that amount.[62] In 1984 at the Bask Classic, a four-year-old mare named Love Potion went to a Thoroughbred horse investor for well over $2 million. Armand Hammer, chairman of Occidental Petroleum Company, became the high bidder at Lasma-run sales in 1981 and 1982. By 1983 he owned forty Arabians under a stud title of Oxy Arabians and purchased mares for $500,000.[63] Movie stars like Shirley MacLaine also bought expensive Arabians.[64]

Lasma was not the only successful dealer in newly imported Arabians, and the Arabian boom reflected more than just sales of Polish-bred horses. Importing horses from Egypt had started up again at about the same time as had importing them from Poland, and American breeders who sought out lines from Egypt began enjoying financial success as well. By the late 1950s the breeding situation in Egypt had changed dramatically from the last days of Ali Pasha Sherif.

The Egyptian government got into horse breeding as early as 1892 to ensure a reliable supply of good cavalry mounts. In 1908 the government established the Royal Agricultural Society to orchestrate the breeding of good horses, and in 1914 the society began to concentrate on breeding Arabians. It started to collect horses from the private studs of Egyptian princes and also acquired stock from the Blunts' Sheykh Obeyd stud. In 1920 the society imported horses from Crabbet. Egyptian breeding at the stud reached new heights after the Second World War. In 1949 the society hired General Pettko von Szandter, a Hungarian refugee who was a renowned horseman, to reorganize breeding programs. Szandter chose to use the stallion Nazeer, bred by the society but not highly esteemed for breeding, as one of the chief sires. Though foaled in 1934, the horse would not sire his first purebred at the stud until 1950, and by 1960 his work had been completed. Szandter oversaw other revolutionary changes in this Egyptian breeding operation before he retired in 1959. In 1952 King Farouk was overthrown, and the Nasser government that ultimately replaced him considered disbanding the stud because of its association with monarchy but allowed it to continue its work under a new name, the Egyptian Agricultural Organization. Szandter oversaw that transition, and he eased the American import of Egyptian Arabians in the late 1950s, which in turn initiated a huge expansion in Egyptian breeding for export to other countries.[65]

Nazeer-bred horses became popular with Americans, and the first to enter the United States arrived in 1958, brought in by Richard Pritzlaff and Don and Judith Forbis. Other breeders followed shortly. In 1959 Douglas Marshall of Gleannloch Farms in Texas saw Witez II at the Hurlbutt ranch and became interested in new lines from other breeding centers in the world. Gleannloch had Arabians of Crabbet and Polish background, but he liked what he called the new Egyptians and began to import them in 1961. The most famous horse he brought to the United States was the slender gray stallion Morafic, son of Nazeer. In 1968 Marshall decided to sell all his domestic-bred horses and concentrate on a pool of straight new Egyptian Arabians. He wanted to keep this type of horse "pure" to what might be defined as a modern "strain."[66] He was not alone. From the beginning, all the importers of the new Nazeer horses argued that the Egyptian horses were the most authentic Arabians.[67] Breeders believed they embodied the purest of Abbas Pasha's breeding, which in turn had sprung from authentic Bedouin breeding. In 1970 Forbis, Pritzlaff, and Marshall formed the Pyramid Society, which intended to preserve this "new" Egyptian type of Arabian, valued for its "old" qualities. The Pyramid Society worked well as a marketing tool for these Egyptian Arabians.

The society held its first auction in 1982, where ten straight Egyptian females sold for an average of $211,000, the highest price being $345,000.[68] Jarrell McCraken of Bentwood Farms is a good example of Egyptian affairs in this market. McCraken had never seen an Arabian before 1969, but when his daughter asked for a horse and he found prices as high as $40,000 for an Egyptian Arabian at Gleannloch, he wanted to learn more about the breed. Intrigued by what he believed was the purity of Egyptian breeding because of its Abbas Pasha base, in 1970 he decided to buy Egyptian Arabians. He had learned a great deal from Richard Pritzlaff and Judith Forbis and quickly became a central player in the Pyramid Society. His concern for purity in breeding led quite naturally to a devotion to pedigree. "My very deep conviction is that the first consideration in purchasing a stallion or mare has to be the pedigree. If the pedigree is not there, it does not matter how good the horse is as an individual." A decent pedigree should give at least five generations, McCraken felt, and he believed that pedigrees were evidence of the genetic components of an individual horse. In 1985 his daughter claimed that 65 percent of their sales went to first-time Arabian buyers,[69] and the prices suggested investment buying. Whereas average transactions in 1980 yielded $144,000, prices were much higher by 1985, when one mare sold for over $2 million and eighteen mares and one stallion averaged $373,000. The sale generated more than $7 million.[70]

Jarrell McCraken made history in 1977 when he syndicated a stallion imported from Egypt, Ibn Moniet El Nefous (sired by Nazeer), for $4 million.[71] Stallion syndication quickly became another form of investment in Arabian horses, and Lasma played a major role in that development too. In 1979 the corporation established a more structured system when it syndicated a stallion for over $6 million.[72] A flood of stallion syndications followed.[73] By the mid-1980s two types of stallion syndication existed: one could be described as nonprofit, in which only shareholders used the stallion for breeding; the other was designed to make a profit by standing the horse at public stud and dividing the income among syndicate owners. Stud fees generally were about $10,000. Syndication tended to limit the stallion's output, raising the value of the foals, which made breeders eager to pay the stud fees.[74]

To create new markets for their customers, Lasma began in 1984 to push for renewed private ownership of expensive stallions, which had decreased with the growth of syndication. The argument rested on the idea that many investors had now acquired expensive mares. The corporation decided to hold a "Lasma Star Stallion Auction" for colts and stallions worth at least $500,000 and aimed its campaign at entrepreneurs who owned more than twenty mares. Lasma touted the financial benefits that would result:

> The justification for ownership of a $1,000,000 stallion to breed as few as ten mares is based on the production of five fillies per year averaging $100,000 in value plus income from standing the stallion at public stud for $5,000 to 50 outside mares. At the end of five years (the usual payment period for a $1,000,000 stallion) the Arabian entrepreneur should have sold, or have retained in inventory, 25 fillies by the stallion representing a total of $2,500,000. An additional $1,250,000 will have been realized in stud service revenue. Without the sale of one colt or syndicating the stallion, the entrepreneur should have realized $3,750,000 on a $1,000,000 investment in just five years. Imagine the potential return in investment should the stallion be syndicated while the entrepreneur is still collecting five $100,000 fillies a year.[75]

The heated market for Arabians made some breeders seek genetics outside centers like Poland and Egypt to serve as investment incentives. Howard Kale Jr., for example, focused on Russian Arabians. The Kales, friends of the LaCroix family, had been to Poland to see horses in the 1960s. Kale thus knew a great deal about the world's Arabians. His interest centered on Russian horses bred at Tersk, and in 1975 he began negotiations for importing horses from that stud. He particularly admired a stallion named Muscat. This horse descended from the Crabbet stallion Naseem (and thus the Polish Skowronek) on his sire's side and from Ofir (sire of Witez II and of Bask's sire) on his dam's.[76] In effect, Kale planned to introduce genetics that had been proved within certain lines in the United States. He managed to import Muscat, who arrived in 1978. Kale built a horse operation that would become Karho Inc. by the mid-1980s, and he intended to market and serve the interests of what could be described as another modern "strain," the pure Russian Arabian. Muscat came into the country owned by a syndicate with twelve members. Within several years, twenty owners held a total of one hundred shares, with each share worth $150,000. Muscat went to stud, originally at a fee of $5,000, but by 1985 that had risen to $25,000.[77] Just as LaCroix capitalized on Bask, so Kale capitalized on Muscat. The stallion became national champion in 1980. His progeny went on to sell well at Karho sales. (The stallion sired 1,005 foals in his lifetime.) At "Karho 83" some $4 million was generated for an average sale per head of $165,000. At "Karho 84" twenty-eight lots sold for $5 million with an average price of $180,000. Muscat daughters averaged $222,000 in 1984 and showed an appreciation of some 42 percent over those sold in 1983.[78]

This ever-growing interest and mushrooming of market activity in the United States triggered the rise of fully orchestrated sales in two of the older and government-organized centers of Arabian horse breeding, Poland and Russia. And, as one would expect, Americans bought most of the horses offered. Certainly the high-priced animals went to Americans. At the Tersk Auction of 1984, for example, the highest-priced horse went to an American for $302,000, but most animals sold for about $75,000.[79] It obviously was worthwhile for an American breeder to buy in Russia and then resell to investors in the United States, even though the importer undertook considerable expense and risk. At the Tersk auction the next year, the average sale price was $58,000, with Kale paying $100,000 apiece for two mares. All the horses sold to non-Russians; buyers came from the Netherlands, Britain, and the United States.[80] Apparently Europeans also bought Arabians at these sales, but European breeders remained heavily influenced by the strength of the American market. A few examples of European breeding efforts illustrate how much the demand for Arabians in the United States affected the fortunes of breeders in Europe.

The stallion Aladdinn played an interesting role in this international world. Bred by Erik Erlandsson and foaled in Sweden in 1975, Aladdinn resulted from the mating of the mare Lalage, foaled in Britain by Lindsay from Polish breeding, to the Polish stallion Nuredinn, who had spent his life as a European circus horse. Erlandsson had been able to locate fine old Polish stallions, some of them the sons of Witraz and Wielki Szlem, working as circus horses in Switzerland and Germany. Most of these stallions, eighteen to twenty-two years old, had been sold because when they were born there was not enough demand for them as breeding stallions. (Bask had managed to escape that fate, even though he had not been used for breeding at Janow.) Polish authorities were able to provide Erlandsson with proper pedigrees for these horses, and he then exported many to the United States. But he especially liked the small stallion Nuredinn and decided to keep him. Erlandsson showed Aladdinn, the colt Nuredinn sired, to LaCroix in 1976, but the doctor displayed no interest. When Erlandsson again showed the young stallion to LaCroix in 1978, the story had changed. LaCroix saw potential in the pure Polish stallion and imported him. It was this horse that LaCroix syndicated in 1979 for over $6 million. The stallion also became American national conformation champion that year, making his syndication easier.[81] By 1999 Aladdinn had sired 1,116 foals.

Another stallion's story also illustrates the connection of European breeders to the American market. In the early 1980s, El Shaklan, a graceful gray stallion bred in Germany from Spanish-Egyptian lines, had established his German breeders on a second stud farm in the United States, had left important progeny in that country (his stud fee there was $10,000), and then had been exported to Brazil. The Om El Arab stud (founded in 1970 and owned by Sigi and Heinz Merz) bred El Shaklan in Germany. His dam, Estopa, a pure Spanish mare, would later be taken to the United States. El Shaklan's sire, bred in Egypt, came from Nazeer lines. El Shaklan launched the Merzes on the international scene, and by 1987 the breeders claimed to have exported 120 horses around the world. They supplied the American market from their base in the United States and the European market from Germany.[82] The Merzes' breeding showed how global the situation had become. Breeding from Spanish and Egyptian lines in Germany, they found they could penetrate the American market. Success in that market led to exports around the world.

The rising boom in Arabian horses in the United States—where investors were prepared to put a great deal of money into the horses—could at least partially be explained by the impact of literature. Arabians became desirable to many people through the novels of Walter Farley, who wrote in the 1940s and 1950s about desert-bred Arabians of extreme beauty and authentic type (notably *The Black Stallion*). His books appealed to many people who were young at that time, and by the 1980s they could realize the dreams Farley's books had stirred in them. One breeder claimed that Farley contributed more to the development of the Arabian than any other single person.[83] Some people became breeders themselves because of Farley's influence, while others tried to recapture childhood memories by investing in Arabians. Farley had a strong impact on Arabian breeding culture in the United States, and he helped inspire the rising interest so evident in the breed after the Second World War. The passion for purity, authenticity, and breeding by Arab standards that Farley had triggered, combined with extensive importing from countries that held ancient Arabian breeding centers, reawakened the whole issue of strains and purity. The three strains defined as Saklawi, Kuhaylan, and Muniqi took on new meaning under these expanded breeding conditions. Authenticity became related to strain as well. American breeders interested in Egyptian Arabians returned to study the work of Carl Raswan on Bedouin breeding.

Raswan, whose name had been Carl Reinhardt Schmidt, was born in

Germany in 1893 and traveled extensively in Arabia during the years around the First World War. He became an American citizen in 1927 and changed his name to that of Raswan, a stallion sired by Skowronek and imported by Kellogg. Schmidt had greatly admired the horse, who died tragically shortly after arriving in California. Carl Raswan wrote about breeding techniques of the Arab people while living in the United States, long after his travels in Arabia. He believed the Arabs had practiced strain breeding and also inbreeding. He argued that in order for modern Arabians to maintain their trueness to authentic type, they should be bred within strain, using inbreeding to stay in strain. Raswan also held certain convictions about the value of the three strains. All breeding should be away from the Muniqi strain, which he claimed was impure as a result of one Bedouin cross with a Turkish stallion in the seventeenth century. A pure or *asil* Arabian had no Muniqi blood. Rawsan stated that Europeans had in the past favored Muniqis because of their racing ability, and that had stimulated their production for the European market. (The Darley Arabian had been a Muniqi.) For their own purposes, the Bedouins did not breed to Muniqi if they could help it.[84] The renewed interest in strains that resulted from Raswan's work made some breeders start to categorize modern Arabians in relation particularly to their origins from Davenport imports, Blunt imports, or Abbas Pasha stock (found in Ali Pasha Sherif breeding). "Blue star" described horses that descended on all sides to these roots and with no Muniqi blood. "Blue list" animals traced to these roots as well, but with 1 to 50 percent breeding of Muniqi. All other Arabians were "general list."[85]

Concern with strain and Bedouin breeding raised interest in Arabians purely descended from the Davenport imports. While the underlying tension in Arabian horse breeding—maintaining original type or improving that type—had attracted breeders to Davenport Arabians from the time they were imported, major breeding establishments devoted to pure Davenports did not exist before the 1950s. These studs practiced Raswan's theories to ensure the continued use of Bedouin breeding methods. Inbreeding within strain and away from Muniqi preserved true type, they maintained, which in turn represented the highest quality. One Davenport breeder put it this way: "The challenge of Davenport breeding has not been in the production of good horses because this has resulted from the automatic biological process." The challenge revolved around finding more people who wanted the really authentic Arabian horse, which had not been bred for either beauty or show.[86] Davenport breeders claimed that perpetuating type brought the highest quality and that beauty had never been a part of the Bedouin horse.

With the new appreciation of different styles in the United States, the uniformity of perceptions on what designated good Arabian type, which had existed with the complete acceptance of Skowronek-Crabbet breeding, broke down. The American Arabian horse world developed internal fractures that defined Arabians in various terms. In a sense, what could be called new "strains" started to grow out of these ideas. Egyptian Arabians remained very different from Polish lines. Crabbet horses still dominated the American Arabian scene in sheer numbers, but they too began to be seen as a subset of the Arabian breed. Western approaches to breeding reflected the absorption of Eastern thinking. The evolution of new "strains" and interest in foreign genetics in the United States only increased American breeders' desire to buy abroad. Investors too were attracted by these lines.

The structure of the American tax system also played a major role in the Arabian boom in the mid-1980s. Valuable Arabian horses could be used as a tax shelter. Arabian horse breeders who knew the best bloodlines around the world found themselves in a position to run a business that had little to do with breeding horses. The issues of the journal *Arabian Horse World* became huge in the 1980s—thousands of pages in some cases—and advertising focused on sales to investors. Less and less material dealt with the practical problems breeders faced. An edition of the journal that contained a thousand pages might devote twenty to breeding. Most information on individual horses concerned what the animals seemed to be worth today and what they would generate tomorrow. All valuable horses were related to the show system by detailed studies of how many wins they or their progeny had gleaned. New "strains" often seemed to be little more than gimmicks devised to keep prices as high as possible. The investing public constantly received information on new merchandise to make money with. Shows played a role in that investing boom.

Arabian showing had developed rapidly and more diversely after the first national show in 1958. By 1970 a complex show structure supported a national system for horse exhibitions. Classes developed for conformation, but also for performance. Horses worked under saddle in park, English pleasure, and western pleasure events and also competed in driving and in native Arabian costume classes. Horses, groomed and clipped to enhance desired characteristics, received extensive training. A need for skilled handlers quickly developed, and the best of them made their living training and showing horses. Even conformation animals had to be handled in specified ways. Show ring results could be used as selling devices, which influenced breeding decisions. Journals on Arabian horses regularly analyzed the record book to reveal which stallion produced the most show champions (mares received attention, but not as frequently or with as much enthusiasm). Two distinct horse types, fancy (for competing in conformation classes) and utility (for performance classes that emphasized gaits), resulted from selective breeding. The growth of Arabian horse racing, well established with large purses by the 1980s (though races had been held as early as 1959), made emphasis on specialized qualities relating to either utility or fancy, but not both together, even more evident.[87]

As the boom reached a peak, many American breeders became unhappy because love of the animals and a wish to breed for quality seemed to have vanished from the Arabian horse world. The Arabian boom exaggerated patterns seen during the Shorthorn and Collie booms. Worries about how money and the demand of a few wealthy people could affect the quality of purebred animals, which had concerned Shorthorn breeders in the 1860s and Collie breeders at the end of the nineteenth century, became evident in the Arabian horse press.

For example, open criticism of the sales developed. The high-priced auctions, one prominent breeder said in 1985, were a joke, and they were also dangerous. "If it weren't for the new people, the breeders couldn't even get rid of their old stuff. But some guy gets conned into paying $300,000–$400,000 for a horse whose truthful value is $10,000–$15,000. Are these people staying in the business? I think it is a real problem." He elaborated: "The fact that we are producing 25,000–30,000 foals a year doesn't mean that the industry is *healthy*. If new tax laws upset these write-offs, we're going to be sitting with thousands of horses nobody wants." The writer believed that the tax situation lay at the bottom of what he called "phony" sales. A horse worth $10,000 gave a tax saving of only $5,000, while one worth $100,000 gave a saving of $50,000. Hence the horse would cost $100,000 to make the saving attractive to the buyer. "We should move away from five or six people getting together and round-robining milliondollar horses and trying to convince the world that that's what they're bringing these days." This trend gave false value to the investment made, and yields could not counter the high costs. "I've seen families broken up," this breeder added, "divorces being created because of the false impression that you can buy a mare for $100,000, take her home, breed her one time and never have to make a second payment."[88]

Some breeders applauded these comments, as a number of letters to the editor of the *Arabian Horse World* make clear. One person stated that the comments "confirmed many suspicions" about the Arabian horse industry: that the high-priced auctions are a joke, and also that "the uneducated attitude of people toward 'straight-something' vrs. Domestic breeding" was detrimental to the production of good horses. This breeder continued, "The only way the Arabian horse will be saved from exploitation is to reduce the tax shelters considerably in order to give the industry back to horsemen and women. Then, maybe showing could be fun again and the horse might enjoy it." Another person wrote that the sales resulted in inflated prices for no good reason and that real breeders were hurt by these sales. "I would like to encourage all breeders of American-bred horses to recognize that we are breeding the finest horses in the world. The focus should not be on pure anything except quality."[89]

Other signs of discontent with the inflated market and the pursuit of investment buyers could be seen in fissions within the organizations that emerged to support the new "strains." Issues of exclusivity and of purity came up as certain groups tried to define the standards they stood for within the heated market. One example could be seen in conflict among the founding members of the Pyramid Society. Established by importers of the new Egyptian, the group had set out to preserve the purity of these Eastern strains. One of the founders of the society, Richard Pritzlaff, had started breeding Arabians of Egyptian blood from a foundation mare named Rabanna (bred in the United States) in 1947 on the advice of Carl Raswan. Pritzlaff and Raswan became fast friends, and Raswan wrote his theories on strain breeding at Pritzlaff's New Mexico ranch. In the 1950s, on the advice of Raswan and Szandter, director of the Egyptian Agricultural Organization, Pritzlaff imported Nazeer-bred stock. Pritzlaff planned to preserve the Arabians he had rather than attempt to improve them.[90] In this way he did not differ from his fellow breeders in the Pyramid Society. But it seemed to him that the others strayed from that mission when breeders of Egyptian Arabians became caught up in the whole investment-driven market for the horses. By 1983 he had become completely disillusioned with the Arabian breeding scene: "Unfortunately, present day Arabian breeders have become more interested in promotion and show winning and in the Arabian horse as a business investment and tax depreciation item than in the Arabian horse himself."[91]

He worried specifically about his fellow Pyramid members, who from his point of view twisted the definition of a pure Egyptian to suit their own purposes and to sell their horses on the heated market. Pritzlaff resigned as an original governing member of that organization because "it had changed and re-defined the definition of a Pyramid straight Egyptian, in an attempt to exclude qualified horses, and [he] believed its effect [was] detrimental to Arabian breeding."[92] Pritzlaff repeatedly commented on what seemed to him to be changing of standards for self-serving reasons. In 1981 he stated, "I am disillusioned, disgusted, and displeased that the Pyramid Society no longer follows and conforms to the purpose for which it was chartered under the law of Texas. It has become a closed society, not recognizing horses which originally qualified, nor recognizing ancestors as roots, discriminating against qualified horses as exceptions, revising and re-revising the original definition."[93]

Pritzlaff had become particularly angry in the late 1970s when the Pyramid Society decided to remove from the "straight" list (meaning horses of pure Egyptian descent as the society defined it) Rabanna (whose blood was now in three-quarters' of Pritzlaff's herd) because she had some Crabbet Egyptian background. Judith Forbis and Jarrell McCracken considered it advisable to label the mare as impure under Pyramid standards. McCracken, as president of the society, stated: "The whole question really has to be whether or not the horses truly qualify. There is enough question existing in the pedigrees to reflect unfavorably upon the integrity of the Pyramid Society, and Egyptian breeding generally, for them to be accepted." "Over 90% of all Pyramid Egyptians have Crabbet Egyptian blood, purchased from Lady Wentworth, from England, in 1920," wrote the incensed Pritzlaff. All of Forbis's Arabians, all Bentwood Arabians, and nearly all Gleannloch Arabians had Crabbet Egyptian blood. "I have had no statement from the Pyramid Society that they have disqualified Rabanna as a Pyramid Straight Egyptian," he noted, but "I hereby challenge President McCracken to prove his statement [that Rabanna was not straight Egyptian] with honest facts and pedigrees."[94] Breeders in general respected Pritzlaff's knowledge and his horses, and Pyramid views did not influence their opinions about the quality or purity of the animals he bred. In 1980 the *Arabian Horse World* commented as follows. "After some forty years in the Arabian horse business, and never participating in horse shows, Richard Pritzlaff is breeding horses that far outclass many of the horses being entered in our shows today."[95]

Changes that would profoundly affect the fortunes of the booming Arabian horse industry were in the wind by late 1985. President Ronald Reagan, who had promised tax reform for six years but had been unable to get a bill through the House of Representatives, had been boxed into an awkward position. The opposition party, the Democrats, had proposed a new tax structure, and it seemed to Reagan that the only way he could get any tax changes through Congress (money bills have to originate in the House of Representatives but must be passed by the Senate as well) was to promise House Republicans that he would pass the bill (the president has the right to veto tax bills) only if it contained depreciation schedules that he and the Republican Party approved of. In December the Ways and Means Committee agreed to a major overhaul of the tax code, and the House of Representatives gave its sanction.

Arabian investors began to be nervous from this point on, but experts on taxes expressed hope in horse journals. Proposed changes seemed minimal and appeared to relate to capital gains and income tax. Depreciation on Arabian horses remained the same. "In fact," one expert reported in the *Arabian Horse World*, "the present status of the legislation, coupled with the promise made by the President to the House Republicans that he would never sign the bill in its present form because he believes it to be unfairly burdensome to business, particularly in the area of depreciation, renders the concerns many breeders and investors have regarding the tax reform unfounded."[96] Corporations like Lasma that had done so well in attracting investors also argued that no danger lay ahead. Gene LaCroix, son of Dr. Eugene LaCroix, claimed early in 1986 that he remained unconcerned about the potential changes in the federal tax code. He believed the Arabian industry could ride out a storm of hesitant investors as well as or better than any industry.[97] Sale results showed that high prices could still be had in 1986. For example, the Lasma Classic averaged $167,000 a lot, the Daughters of Bask Sale brought $179,000 a lot, and the Oxy Masterpiece Sale (for Armand Hammer's Oxy Arabians) averaged $99,000.[98]

But dramatic changes in the tax structure lay just around the corner. Some agreement was worked out between Reagan and the Republicans, and it brought an entirely new concept. The Tax Reform Act, signed by Reagan on 22 October 1986, introduced the idea of passive and active investment. The new act brought major changes to the tax structure—the most significant since 1942, when income tax had been more broadly implemented. Tax imputation after 1986 would be based on categorizing the way money was invested. The act of 1986 limited depreciation allowances and also introduced the passive activity rules, which made it harder to apply current passive losses against other types of income. For those who took an inactive or passive role in the object invested in, tax advantages would be reduced. Active investors could benefit most from sheltering tax losses. The Tax Reform Act, then, changed the situation for the Arabian horse industry by reducing tax shelters for people investing in the animals for tax benefits. Basically it eliminated the tax advantages of owning horses as a passive investor or through passive activity (an investment partnership).[99] The implications of these new regulations did not seem obvious to many in early 1987, though. It appeared that the problem hinged on how one defined investing in Arabian horses, and some thought that by extravagant advertising and other schemes the Internal Revenue Service could be made to see Arabian horses as an active investment. As the *Arabian Horse World* told its readers, "The fact of the matter is that the American horse industry fared extraordinary well under the new tax law, particularly when compared with other interests which have formidable lobbies in Washington, D.C."[100] One breeder announced in 1987 that the tax reform made Arabians one of the best investments one could make. On top of that, this breeder argued, "Few other businesses offer financial dividends with personal fulfillment and family fun."[101]

The letdown came quickly. The bloom on the rose began to fade first within the United States. In 1987, for example, the Merzes felt the turn from their base in the United States when they attempted, because of divorce, to sell their very best horses.[102] By 1988 sales in the United States had plummeted and averaged $10,000 to $25,000 per lot.[103] The *Arabian Horse World* reflected these changing conditions: it had shrunk to about one-quarter of the size it had been three years earlier. As the bubble burst, a number of the big operations began to go bankrupt. The fate of Jarrell McCracken's Bentwood Farms is one example. "When the Bentwood Farm bankruptcy became public [in 1987]," the *Arabian Horse World* noted, "the prospect of some 1,000 Egyptian Arabians hitting the marketplace sent a collective shudder through the Egyptian breeding community. The prevailing public perception was that a sale of so many horses would severely hurt the Egyptian Arabian market, a concern shared by Sotheby's, as well as First Interstate Bank of Arizona, the creditor." The herd was culled so that only a limited number could be auctioned. The Sotheby's sale in July 1988 brought some four hundred people, and eighty-nine horses—all mares—sold for an average price of $22,000.[104]

Lasma Corporation also experienced difficulties. Reorganization resulted in the liquidation of holdings in Arizona, the formation of a new company under the name LaCroix Ltd., headquartered in Texas, and the eventual downfall of the operation known as Lasma East in Kentucky. By 1987 the LaCroix family had introduced a new type of auction—the herd liquidation sale.[105] Preparing for the 1988 sales, they addressed the Arabian horse community in some detail about the industry's difficulties.

The past several years have taught many of us a valuable lesson: *nothing goes up forever,* including Arabian horses, stocks and oil. Importantly, however, *nothing goes down forever,* and realizing this fact brings the true value of the Arabian horse back into focus. Those of us who

The logo of the Lasma Corporation. The image of Bask portrays the fantasy dream of the pure desert Arabian, a fiery steed of immense purity. Bask prances like a Bedouin warhorse. The painting mimics the style of the German artist Adolf Schreyer (1828–99). Breeders greatly admired Schreyer's work and related easily to the characteristics the artist ascribed to Arabians in his paintings. The symbol of the Lasma Corporation thus unites fantasy and beauty with the money-making power of a particular horse. The stallion personified these features that had become embedded in Arabian horse breeding in the United States by the 1980s. The logo appeared many times in advertisements for Lasma Corporation in the 1980s.

really appreciate this beautiful animal and its intrinsic values now, will benefit tremendously in the years to come. . . . The market correction of the past two years, redirects the serious breeder towards improving the desired individual qualities of his stock, and as a result we will be breeding better Arabian horses. We will also direct our marketing efforts towards new people, emphasizing the Arabians themselves, rather than strictly the financial aspects. As a result we will again dramatically enlarge the pool of potential owners who can afford and enjoy the real values of the Arabian. As in the seventies, when people received tremendous satisfaction at more affordable prices, the demand for Arabians will increase.

Lasma expanded as follows:

For the sales in 1988 and 89, we are *intentionally* selling a lot of horses in our sales in 1988. This may naturally drop averages. Buyers will be purchasing better horses for less money than they did in the mid 70s. Therefore for less money, they will receive greater value and greater satisfaction. Those who buy in 1988, and those who wished they had bought in 1988, will be buyers in 1989.[106]

Lasma returned to the idea that a horse should be bought for its quality and that the market should be aimed at breeders, not investors. The huge LaCroix operations, however, would not survive the decline of the hot Arabian market.

Breeders of Arabians outside the United States did not seem to recognize how severely American tax reform would affect the international industry. An example can be seen in Canada. The Cole family of Stonebridge Farms, near London, Ontario, had been breeding Egyptian Arabians since the 1970s and had been members of the Pyramid Society. Their main stallion, Dalul, a son of Morafic, had been bought from Marshall of Gleannloch Farms in Texas. Stonebridge began in the 1980s to offer investment potential to Canadians on the faith that demand for Arabians in the United States would remain strong, because that demand kept world prices so high. Their product was pure Egyptian Arabians. By 1987 the Coles had stepped up their investment schemes for Arabians. Mark Cole explained the position of Stonebridge: "Basically, as I see it, Stonebridge is no longer a conventional Arabian horse operation. We are the creators of investment vehicles through which investors can participate in the Arabian industry on a modest scale. With only disposable income at risk and horses as the equity medium, our limited partnerships act as a show window for the entire industry." By 1987 about six thousand people had invested $20 million. Most investors were Canadian, but Cole hoped to penetrate an American market, which by this time was in decline, with a new Wall Street offering. The most common type of investment was the buying by Canadians of a public limited partnership at $2,500 a unit.[107] The *Arabian Horse World* watched this Canadian investment activity with some interest. "If the sizzle is out of the breed, nobody bothered to tell the Coles, Stonebridge Farm, and their shareholders," the journal noted in 1987.[108] By November 1988 Stonebridge Inc., was listed on the Toronto Stock Exchange, with 1.7 million shares for sale at $10 a share.[109] But Stonebridge's fortunes and those of its investors in Canada would not survive the collapse of the American market. The company went bankrupt.

Some European breeding centers also did not recognize how much American tax reform would influence the international industry. In 1988 Tersk offered eighty-seven head. Only eighteen horses sold. The highest-priced animals went for $36,000, and they "would have been good at 10 times that price" a few years ago, the *Arabian Horse World* reported. Americans did not come. One European buyer bought eleven of the eighteen sold, and only three buyers were present. "In keeping with the principle that everything takes a little longer in Russia, it seems that the news of the bad market around the world has not penetrated as far as Tersk. If it has, one can only wonder at the rampant optimism that made the Russians offer a staggering total of 87 horses, more than ever before," the journal noted.[110] By 1988 breeders in Egypt recognized that the American market affected the value of their horses. Top prices had dropped from $70,000 to $50,000.[111]

By 1988 the international Arabian industry had faltered. It would not be long before that fact became apparent to breeders all over the world. Registration numbers for Arabians in the studbook of the Arabian Horse Registry of America over its history of recording reveal just how hot those times had been, showing how accelerated growth had been in the boom years. From initial registration of 71 horses at the end of 1909, the total increased to 14,000 entries by 1958 and 35,000 by 1963. This increased to 100,000 by 1973, 200,000 by 1980, 300,000 by 1984, and 400,000 by 1988. Growth slowed in the 1990s, and the number reached 500,000 by 1994.[112] The Arabian market struggled to recover equilibrium. The high-priced sales of the mid-1980s could at least partially be blamed for some ill effects on the Arabian horse breeding world, felt throughout the 1990s.[113]

The Arabian horse world would, however, be troubled in the late twentieth century by issues other than boom or bust markets. As the British-American axis involved other countries in the trade, the issue of standards and pedigree acceptance between countries—and the concepts of purity and authenticity—became even more critical. All countries found themselves drawn into the problems of purity, authenticity, and even quality.

World trade in animals registered in different national public records brought various registry systems into conflict because of the standards maintained by the Arabian Horse Registry of America. American breeders set world standards by accepting or rejecting foreign pedigrees. Breeders in the United States and in other countries began to question the hegemony of the Arabian Horse Registry of America and the control it had over questions of purity and quality as defined in pedigrees.

# Six

# THE ARABIAN HORSE REGISTRY
# OF AMERICA: PRESERVING PURITY

"It is the purpose and responsibility of the Arabian Horse Registry of America, Inc. to preserve the integrity of the blood of the purebred Arabian horse. The Registry attempts to honor this commitment by registering purebred Arabian horses, maintaining a permanent record and engaging in activities which aid, promote, and foster the preservation and betterment of purebred Arabian horses and the Arabian breed."[1] This is the mission statement of the Arabian Horse Registry of America (AHRA). Pedigree standards set by the AHRA, in order to preserve purity, were to have a profound effect on the world's breeding of Arabians because of the significance of the American market. This chapter explores the way American pedigree standards developed and assesses their influence on international standards and the dynamics of a global Arabian industry.

World Arabian breeding culture, even if it did represent a fusion of Eastern and Western thinking, took shape over the years in relation to the way the AHRA defined authenticated purity, because the policies of the American registry in effect governed the world's markets. From the beginning, the AHRA defined a pure Arabian as one that came entirely from Bedouin-bred stock, and the organization would pedigree only horses that fit that description. While that resolution would not change, what was believed to authenticate such purity would. Therein lay the problem. In 1908 the AHRA accepted without question Arabian horses from the Jockey Club studbooks of Britain, France, Australia, and the United States (and this meant Canadian Arabians too, because of the close relationship of the Canadian racehorse industry to the American one). Such horses were defined as "pure," that is, as descended entirely from desert-bred horses. Animals recorded in these books qualified for what was called nondiscretionary status and were automatically eligible for American pedigrees. Horses not emanating from those sources fell into the discretionary category; they needed supporting documentation to prove their purity, and the directors of the association had the power to accept or reject such documentation.

As time passed, the AHRA put other public registries on the nondiscretionary list, so horses recorded in them came to be automatically accepted as well. In 1937, for example, the public studbook of the Polish Arab Horse Breeding Society (founded in 1926) became part of the nondiscretionary list. Many horses bred at the new government centers in Poland descended from stock that originated at Slawata and Antoniny, and private records from these studs formed the basis of pedigrees in the Polish Arab Horse Breeding Society's record book.[2] The purity of Polish Arabians' descent from foundation desert-bred stock could therefore be authenticated only by the records of these old princely studs. This problem would haunt the AHRA. In 1938 the studbook of the Arab Horse Society of England (established in 1919) was put on the nondiscretionary list. The General Stud Book (GSB) continued to register Arabians in Britain until 1965, however, and many considered it the more prestigious book. In the 1930s many people suspected that inferior horses had been pedigreed in the Arab Horse Society's book.[3] Neither record, then, was problem-free. Directors owned most Polish and British horses imported in those two years and registered in either of these books.[4]

After that time the AHRA's directors increased their power to confer American pedigree status or to decide what qualified as pure desert descent. Starting in 1950 the association decreed that horses outside the accepted list needed more that an assessment by the directors of their pedigrees and country of origin to guarantee their descent from pure desert-bred horses. Directors, or their representatives, also visually evaluated the quality of incoming stock. The authorities of the AHRA scrutinized candidates for registry and decided whether the horses looked like desirable Arabian types.[5]

From the beginning, the AHRA found it harder to regulate imports from Eastern countries than from Europe, because avenues of entry to the United States from the Middle East were complicated. Over the years pedigree keeping in the East also shifted in a complex way between private and public recording. The story of the AHRA's changing attitude to Egyptian Arabians, in relation to import patterns and to historical pedigree keeping in Egypt, illustrates these difficulties. Although most Egyptian Arabians arriving in the United States before 1918 had originated from private Egyptian studs, the animals qualified for nondiscretionary entry. The horses came by way of England with GSB pedigrees (usually through sales from the Crabbet Arabian Stud). Under these conditions, their pedigree status in Egypt did not matter. The problem of pedigrees for Egyptian horses arose when the animals came directly from Egypt, a route that became more frequent in the second half of the twentieth century. Historical pedigree keeping in Egypt became critical at that point.

After 1914 Egyptian horses could be pedigreed not just by private studs, but also by the new government operation in Egypt, which bred Arabians. The government stud maintained pedigrees, and even though these records pertained to only one operation, the move could be described as a deviation from standard Eastern recording practices. The records reflected a public, government-controlled breeding program, not a private, personal one. Private breeding centers continued to exist in Egypt, and these maintained their own records. (No association of Arabian horse breeders as a whole existed in Egypt until 1985.) The multiplicity of stud records and the lack of a breeder organization might have made it difficult for the AHRA to establish clear-cut qualifications for Egyptian Arabians.

Increased American demand for Egyptian Arabians, evident by the mid-twentieth century, and traditional acceptance of Egyptian Arabians on the nondiscretionary list (because of their GSB pedigrees) led the AHRA to recognize Egypt in 1964 as a nondiscretionary source. One other factor probably played a role. Egyptian horses in general became significant because American breeders believed they often carried special authenticity from a connection to Abbas Pasha breeding, which in turn took the stock directly back to Bedouin desert-bred lines. From 1914 Egypt's Royal Agricultural Society had focused on preserving the lines of Ali Pasha Sherif and, through those lines, the blood of Abbas Pasha's horses. The 1952 change in the organization (it became the Egyptian Agricultural Organization) did not alter these breeding goals.[6] While between 1964 and 1972 the AHRA accepted Egyptian Arabians from the government stud as nondiscretionary, the registry also made horses from private studs eligible for pedigrees. Some breeders believed it remained unclear under which category—discretionary or nondiscretionary—horses from private studs received American pedigree status. Recording practices in Egypt might have complicated the issue, but many believed that directors—who again did most of the importing of Egyptian horses in these years—deliberately kept the matter imprecise to serve their own interests. Although by 1972 wording for nondiscretionary Egyptian Arabians had changed on the AHRA's import regulations to suggest that private studs qualified, no clear definition of what made an Egyptian Arabian automatically acceptable for American registration had been provided.[7]

Importing by breeders from other countries where horses qualified only for the discretionary list could make the AHRA coincidently change its regulations. The status of Russian Arabians, for example, shifted when Howard Kale imported Muscat in 1978. Before that year horses bred in Russia, even though they were registered in a public studbook, could receive pedigree status in the United States only if they had passed through Poland or Britain and been pedigreed there. The AHRA accepted such pedigrees without needing the discretionary authority of the directors. If the same animals came from Russia to the United States with only Russian pedigrees, they could not be papered as pure Arabians even if they carried only Crabbet and Polish bloodlines. In the 1960s the loophole worried the authorities of the AHRA. The entry of Russian horses through the British studbook even led the president to consider rescinding the special status that the British book's pedigrees conferred for admittance to the American book.[8] British pedigrees had been the parent base of American Arabian pedigrees; the GSB records had guaranteed purity from the late 1870s. The issue of AHRA objectivity became increasingly contentious when international tensions over pedigrees erupted at the end of the twentieth century.

At the same time that standards of pedigree under the AHRA began shifting over the mid-twentieth century, a move to organize Arabian horse breeders on a global level was under way. In 1967 the Arab Horse Society of Britain called a meeting to discuss forming a world organization for Arabian horse breeders to deal with shared international problems. Representatives from fifteen countries attended the meeting called by the Arab Horse Society of Britain, and they created the World Arabian Horse Organization (WAHO). An American breeder who attended, Jay Stream, was soon to be of great importance to the international Arabian horse breeding fraternity. Stream, a land developer who grew up on an Iowa farm and bought his first Arabian in 1960, had studied pedigrees and traveled throughout the world looking at breeding programs. He imported Egyptian horses, admired Polish horses, and knew Skowronek-bred Arabians.[9] His interest in the global situation developed rapidly after 1967.

In 1970 twenty nations attended another meeting in London and set up a steering committee under Stream to create a constitution for the organization. This work took two years, and at the next meeting in Spain, the twenty-seven countries attending ratified it. Countries had five years to apply for membership and another five years to get their studbooks in order. Studbooks needed to have a five-generation pedigree system, except in some Middle Eastern countries where written pedigrees did not exist.[10]

The organization had agreed to accept two types of members. The first, called "big M's," represented the registration authorities of the various countries. The second, called "small m's," individual associate members, comprised individuals who supported WAHO.[11] Thus, by 1972 WAHO had established an operating structure and also approached the problem of international standards.

Stream's interest in international standards, and the dichotomies they could present, increased through direct contact with Spanish horses while he was at the 1972 meeting. He liked An Malik, a Spanish champion stallion. The AHRA would not accept this horse for registration. The historical background of Spanish breeding and An Malik's ancestry show why the AHRA had trouble authenticating the purity of this stallion and many other Spanish horses. Arabians had been bred in Spain for hundreds of years. In fact a national and public studbook for Arabians existed in Spain as early in 1848, though no pedigrees were published until 1885. The Spanish government carefully edited pedigrees of the first studbook, because Arab occupation of Spain for so many years meant that some confusion could exist between Oriental Barbs, Andalusians, and Arabians. Until the end of the nineteenth century there were few private studs in Spain. Early in the twentieth century, however, important new breeders appeared, and in the 1920s and 1930s they imported Crabbet Arabians. The Duke of Veragua, a particularly important breeder, brought in Skowronek daughters. When the Spanish Civil War disrupted this famous stud, the national cavalry managed to move the horses to safety. Most of the mares had foals at the time, and although everyone knew they were purebred Arabians, no one could be sure of the correct sires and dams.[12] Many important Spanish Arabians recorded in the Spanish national book by the 1960s descended from Veragua mares. An Malik was one of these.

Stream convinced WAHO at the Spanish meeting to accept Spanish horses from Veragua background as purebred and to respect the standards of the Spanish book. Interest in the old Skowronek-Crabbet lines made many people agree with Stream that the horses should be accepted as pure in spite of the incomplete records. No one had questioned the purity of the Veragua mares dispersed by the Spanish Civil War. The problem was the difficulty of pedigreeing the animals properly under Western record-keeping structures because of their unknown parentage. Stream also managed to get the AHRA to accept these horses for registration. Arabians known to have descended from the Duke of Veragua's stud received pedigrees under the name "Veragua." (Stream bought An Malik for $10,000.[13] The horse would stand in the United States for a stud fee of that amount.[14]

In 1983 Stream held a sale of An Malik daughters; Tom Chauncey became the highest bidder at $250,000.)[15]

The Veragua problem brought into the open the question of conflicting standards and showed how they skewed the market. Because the AHRA controlled what horses could enter the United States as purebred, and because American trade also drove the international market, registries in countries where the AHRA would not accept pedigrees could not command the respect of the world Arabian community. Under these conditions, quality did not dictate acceptance, in the United States or ultimately in that wider community. A new strategy emerged in WAHO out of the controversy on how to deal with clashing standards. If registries in the world could agree on a definition of purebred Arabians, then notions of purity would no longer differ between countries. WAHO members agreed to try to establish a world definition for a purebred Arabian horse, and Stream set out to see that the organization did so. He devoted the next two years to the project.

For the first time, under Stream's guidance Arabian horse breeders from around the world worked together to decide what it meant to call a horse an Arabian. Members of the executive committee of WAHO met frequently with delegates of member nations to keep in touch with what the world community thought. By 1974 they had hammered out a definition, but not without difficulties. A new journal, the *Arabian*, established in Britain to serve the international Arabian horse breeding community, described problems the breeders encountered when they faced this task. The *Arabian* had this to say about the journey that led to the 1974 WAHO world definition of the Arabian horse in Malmö, Sweden: First, origins of first Arabians found in all studbooks had to be agreed on—not too difficult a task. The *Arabian* explained that everyone accepted that these origins "were, first, the horses bred by the horse-breeding tribes of the Arabian Desert, and secondly those registered in the older and better established Stud Books of the world. These horses and their descendents, it was felt, must account for all the purebreds which were already registered in Stud Books or Registers. Since several countries had expressed doubts about whether there were in fact any purebred Arabians in Arabia by the time of the Second World War, it seemed logical to add a time limit to the so-called primary origin and say that it would be regarded as an origin of purebred only up to" that war.[16] But difficulties entrenched in the Eastern and Western heritage of Arabian horse breeding, and in attitudes toward both, immediately set in. The *Arabian* continued, in a somewhat confused fashion:

But then, since all the "desert-bred" horses with which the registries of the world were concerned were already listed in one or other registrar or stud book, there seemed no point in keeping the horses of the desert of Arabia as an origin separate from horses bred there which had been registered in a western type Stud Book. This was a logical but at first emotionally unacceptable step. After all the Desert was "The Origin" and so why not trace back to it whenever possible? The answer was that the Desert, even when carefully defined geographically and in terms of tribal migrations, was a nebulous and slightly unreal concept, even though it was an essential part of the romanticism with which many people approached the breeding of Arabian horses. What was real, however, and had been used as the basis of pure Arabian breeding in every country outside the Middle East itself, was the population of horses which came from the desert at a time when they were still bred in the traditional way (this, so to say, guaranteed or constituted their purity) and had been registered in the first established Stud Books.[17]

Certain studbooks, then, contained the names of the original Arabians to which all "pure" horses traced, but newer studbooks could be said to be reliable in spite of the lack of recording of original animals because they used the older ones as a starting point. Registries set up in Canada, Australia, and South Africa—all established about 1960 and based on the older books—illustrated this characteristic. Defining a horse as purebred solely on its relatedness to documented desert origin, then, did not seem practical.

Another major problem resulted because some horses in every studbook—even the older ones—could be found not to trace accurately on all sides to what could be described as authenticated "pure" Arabians. "No Stud Book was free from horses which could be shown to be lacking in proof of descent from none but Arabian horses," the *Arabian* noted. As a result, "a Definition based on approved origins tended to be 'exclusive' and in some cases unjustly so, and, if pursued to its logical conclusion, would entail the elimination of horses at present accepted as pure in one or another Stud Book." The final definition accepted at Malmö would be, "a purebred Arabian horse is one which appeared in any purebred Arabian Stud Book or Registry listed by WAHO as acceptable," and the books of the United States, Britain, Egypt, Poland, Australia, Denmark, Germany (Marbach), the Netherlands, Canada, New Zealand, South Africa, Spain, and Sweden were accepted immediately. Members hoped that by 1979 all studbooks in the world would be recognized.[18]

The executive committee then looked into how to qualify the status of certain horses that were registered in various studbooks. Pedigrees of animals around the world contained problem horses with known questionable backgrounds that could not receive pedigree status from some registries. Since these horses could be described as "impure," so could their descendants. One horse in this contentious position was the stallion Kurdo III, who came to Argentina from Germany in 1910. The *Arabian* explained:

[Kurdo III] was used extensively in pure Arabian breeding in Argentina and had influence on breeding in several European countries to which his descendents had been exported. He had never been thought of except as a purebred Arabian, but it was discovered that his pedigree could be traced back a very long way, at least in certain lines, and that in one of these lines there appeared to be an English Thoroughbred mare [30 Maria] registered in vol. 5 of the GSB and born in 1842. Her pedigree could in turn be traced back to many presumed Arabian horses which were the foundation stock of the English Thoroughbred. [Kurdo III] was accepted as a pure Arabian because this line to English Thoroughbred came in over 60 years before he was born and was itself not completely non-Arabian and no other addition of non-Arabian blood was known. A further consideration was that this horse had bred on well and descendents of his which had been seen by members of the Executive Committee showed both type and style.[19]

It might be pointed out that the inclusion of the mare in the GSB did not in itself prove she was not purebred Arabian. Clearly, though, one could not prove with certainly that she was. WAHO decided to accept Kurdo III as purebred, and therefore all of his descendants. The organization saw the issue as one of practicality.

The work of WAHO and of Jay Stream did much to create a system for world standards. Easier movement of horses between countries followed immediately, and that in itself fed importing to the United States in particular; when Americans imported, other countries did so as well in order to produce stock that ultimately would go to the American market. It also fed the export market out of the United States. "The biggest contribution that WAHO has made is to standardize the stud books to some extent and to require enough recording on import-exports so that we know in fact that the horse actually coming into the country matches his papers," said Stream in 1985.[20] More studbooks achieved recognition under the WAHO definition of what made an Arabian pure, and by 1986, thirty-seven member countries had their record books approved.[21] By 1992 that number had grown to forty-seven, in spite of the difficulties that had arisen in the American market. By 1999 that number was fifty-four.

The growth of WAHO, however, did not bring an end to friction in the international Arabian breeding world. Trouble had been brewing for some time, because the largest registry in the world hesitated to adopt WAHO's definition of purity in the Arabian horse. By the 1980s the Arabian Horse Registry of America, a member of WAHO from the beginning, began telling the world organization that although it intended to accept WAHO's definition, it needed time to do so. At the same time, the AHRA would not accept some registries recognized by WAHO, such as Saudi Arabia (passed by WAHO in 1986) and Syria (accepted by WAHO in 1990). Of WAHO member countries in 1991, the United States remained the only one that had not adopted the WAHO definition as the basis for import regulations. The American registry rejected all pedigrees from 35 percent of WAHO countries, even though delegates from the AHRA had voted in favor of all the policies these countries followed.[22] The AHRA stood by its claim that it protected the purity of the breed, and it defined purity as being descended entirely from desert-bred stock.

The issue came to a head over descendants of Kurdo III. In the 1990s the AHRA refused to accept the pedigrees of certain Argentine horses that had descended from the controversial stallion. The AHRA claimed that its not accepting Kurdo III protected the purity of the breed. The registry recognized as purebred only horses that could be traced to the desert on all sides. The authorities of the American registry also argued that national sovereignty was at stake. It would be American breeders and only American breeders who would decide what horses qualified as purebred in the United States. At the 1996 meeting of WAHO at Abu Dhabi, members agreed that the AHRA would be suspended in February 1997 unless it agreed to accept WAHO's definition of a purebred Arabian. Thus the largest single registry in the world would be suspended by a world organization. WAHO countries, in turn, would no longer accept American-registered horses for import to their nations.

The conflict between the AHRA and WAHO intensified the arguments about what it meant to be a purebred Arabian horse. WAHO took the following position. First, the organization went to great lengths to show that the American registry contained many pedigrees that did not trace on all sides to the desert. To begin with, WAHO pointed out, "Egypt is in North Africa, not in Arabia." The organization then elaborated: "By no stretch of the imagination can Syria, Poland, Turkey or Spain be considered 'the deserts of Arabia.' And yet many horses registered in AHRA trace back to horses whose known ancestry stops in these countries. This does not make them 'bad' or 'impure' Arabians but they demonstrably do not 'trace back in every line to the deserts of Arabia.'"[23] WAHO next examined pedigrees to prove known lack of desert descent in AHRA-recorded stock. It was not hard to do, as is clear from the example of Skowronek's recorded ancestry. The stallion's background showed that he could not be traced, through complete and accurate documentation, to desert breeding. WAHO claimed that authorities of the AHRA knew of Skowronek's undocumented background as early as 1926, but that did not stop importing of the stallion's progeny.

Skowronek, the great breeding stallion of the Crabbet Arabian Stud, had been bred by Potocki at the Antoniny Stud in Poland, foaled in 1908, and later imported to England by an American. Shortly after his import, Lady Wentworth acquired the horse. To register him in the GSB, Wentworth had to go to his breeder for papers authenticating his past. The Potocki family supplied the following information. Skowronek's sire, the gray stallion Ibrahim, had been brought to Poland in 1907 by Potocki. The animal, said to have been purchased in Odessa by a Potocki agent who had brought the horse from the Orient by way of Constantinople, was documented as sired by Heifer out of Lafitte.[24] Skowronek's dam, Jaskulka, had been bred in Poland from Polish stock. Registration in the GSB on the strength of this information certified Skowronek as purebred, a fact that gave his progeny automatic access to the Arabian Horse Registry of America, in turn implying that all his ancestors could be traced to desert breeding.

Even though Skowronek held pedigree status in the GSB, Wentworth decided to invent a more glorious ancestry for his sire, Ibrahim, than "sired by Heifer out of Lafitte" and for his dam, Jaskulka, than merely "Polish breeding." Skowronek reminded Wentworth of the ethereal horses her parents had found in Egypt at Ali Pasha Sherif's stud in its final days. Because Ali Pasha Sherif was credited with preserving Abbas Pasha lines, Wentworth connected Skowronek with Abbas Pasha breeding, the fount of all Arabian horse authenticity. "Skowronek," she wrote, "though foaled in Poland, returns by both sire and dam to the old Abbas-Crabbet strains." Wentworth used Lady Anne's diary to prove that the sire and dam of Ibrahim could be traced to strains of Abbas Pasha, and also to the same strains that the Blunts had acquired. Skowronek's dam, Wentworth claimed, was known to be pure Arabian, and she too, as evinced by Lady Anne's work, descended from Abbas stock.[25] No proof, of course, exists that either of Skowronek's parents descended from Abbas Pasha lines. The Potocki family reconfirmed the background of Skowronek's sire. The son of Skowronek's breeder, Joseph Potocki Jr., prepared a statement some years later making it clear that Ibrahim's pedigree could not go beyond the names given for his sire and dam and therefore could not be traced to Abbas Pasha breeding. "I have jotted down these memories of Skowronek and Antoniny," Potocki wrote, "hoping they may interest those who possess his offspring in their breeding establishments. I am anxious to remove any doubts which may have arisen concerning his true and authentic pedigree."[26]

Real evidence exists not only that Skowronek's dam, Jaskulka, did not come from Abbas Pasha lines but that she was not even purebred. A history of the Sanguszko family stud, written in 1876 and published in the United States in the 1960s, stated that all imports of Arabians by that family before 1818–19 had been stallions. Purebred mares did not come into the Arabian breeding herd until later. Stallions imported before 1818 bred local mares and, over time, cross-bred mares that resulted from these unions. Thus all horses that could be traced to the pre-1819 Sanguszko stud records (as could Jaskulka) should not be described as purebred. In effect, they were bred up.[27] Equally damning evidence about the impurity of Jaskulka emerged through comments made in 1900 by Prince Roman Sanguszko (1837–1917) about horses at Slawata (which was, of course, connected to Antoniny). The prince claimed that in 1900 the operation at Slawata had no purebred Arabians and that all the horses traced at least in part to local horses. His animals did not, he stated, descend on all lines from desert-bred imports. Later it would be argued that some of Jaskulka's ancestors listed in the family records came from specifically known non-Arab lines of the Slawata stud.[28]

In WAHO's Publication 21, released in January 1998, the organization compared the relative known purity of Skowronek with that of Kurdo III. Skowronek, in the sixth generation of his pedigree, had nine out of sixty-four ancestors that had, or could have had, impure Arabian blood. Kurdo III had one—30 Maria (and she only possibly). In the seventh generation, Kurdo III had three possibly impure Arabians and Skowronek had sixteen. "In fact," WAHO stated, "if we count every recorded Arab ancestor in both pedigrees, Kurdo III has more documented Arab blood (98% compared to 93%) even without counting the Arabian ancestors of 30 Maria." WAHO analyzed other pedigrees and pursued the issue with some glee. Many present-day Polish American Arabians descended from pre–First

World War private Polish studs that had kept records of impure stock bred to pure Arabians over the nineteenth century. Certain horses of this background figured in the pedigrees of Arabians bred at the state studs after the Second World War. Both Bask and Witez II were descendants of the impure or unknown nineteenth-century stock. Under these conditions all horses tracing to Bask (some 25 percent of AHRA pedigrees), or to Witez II (some 33 percent) came from known possible impurity and did not trace to the desert by documentation on all sides.[29]

The second issue that WAHO brought up was the AHRA's national sovereignty stand. "Of course, there is not and cannot be any issue of national sovereignty involved, for the simple reason that the AHRA is not a sovereign nation: it is not a nation at all," WAHO pointed out. "It has no sovereign powers at all. It is not the United States of America. It cannot claim the respect or loyalty Americans show their country. It is a private corporation under the complete control of a small and self-sustaining group of people."[30] WAHO contended that the AHRA had always looked after the interests of particular breeders at particular times when they wanted to import new breeding stock. In fact Davenport had been motivated to found the AHRA so he could pedigree his imports in the United States. When the Polish studbook gained acceptance by the AHRA in 1937–38, at least some horses, known not to trace to the desert, had been brought in from Poland by AHRA directors. The nonacceptance of Russian horses by the AHRA in the 1960s had been nothing more than a political move, WAHO claimed, and the move to stop their inflow into the United States through Britain or Poland remained part of that mentality. It was not concern for purity or accuracy of pedigrees that lay behind this thinking. (One could add that many Russian horses might be described as "impure" through Skowronek and Naseem.) In fact, WAHO argued, the AHRA wanted to force other countries to accept their "impure" stock—descendants of Bask, for example—at the same time that it would not recognize "nonpure" stock from those nations.[31]

The Arabian Horse Registry of America had its answers to the criticism. The directors of WAHO, the president of the AHRA said, "apparently believe, that generations of breeding will make impure blood disappear and cite the case of thousands of horses in South America. WAHO believes that these horses are sufficiently pure enough to quality as purebred. The Registry disagrees with this position."[32] "The Registry has made the right decision to protect the purity of the Arabian breed in America, and that integrity will prevail in the long run over political expediency," AHRA emphasized.[33] It was, the registry repeated, also an issue of power. "In essence, American breeders are being told by WAHO they must surrender control of the integrity of the Arabian horse in the US to an organization that demands registration of proven impure horses and whose rules and regulations change at a whim." The AHRA took on the issue of having impure horses in their studbook as well. "Unfortunately, no stud book is immune from finger pointing and rumors of impurity. However, there is considerable difference between registering horses that are known to be impure and registering horses that are questioned, [when] proof that they are not purebred [is] lacking."[34]

All registries would be drawn into the controversy as WAHO and the AHRA squared off. Canada, for example, as a member of WAHO, found that expulsion of the United States from that organization made Canada's position difficult. The historical background of Canadian Arabian breeding shows clearly how closely the Canadian situation had been linked to the American one. Until 1958 Canadian breeders functioned entirely as a satellite of the United States with respect to Arabian horse matters. Canadian horses received pedigrees from the Arabian Horse Registry of America, although recording could be done in Canada with the Canadian Livestock Records Corporation, through the General Stud and Herd Book (which registered a number of breeds). Canadian breeders formed an association in 1958 and established the Canadian Arabian Horse Registry (CAHR), under the Live Stock Pedigree Act. Canadian breeders could still register in the American book as well, but until 1976 the AHRA would not allow entry into that book based on Canadian pedigrees alone.[35] In other words, horses with Canadian pedigrees qualified only for the discretionary list. Canadian breeders, then, had to register in both books in order to function as purebred breeders with respect to entry requirements for shows and to potential markets in both countries. When the American registry granted nondiscretionary status to Canadian pedigrees for American recording purposes in 1976, Canadian breeders achieved some national independence. The Canadian national show, initiated in the late 1950s, had became part of prestigious showing for American breeders by the 1980s, and valuable horses regularly went north to compete for the Canadian part of the North American Show crown. Pedigree issues, however, remained entangled with entry requirements, thereby ensuring that breeding activities in Canada never become completely independent from those in the United States.

In May 1998 the Canadian association notified WAHO that it wanted to resign. Under the WAHO constitution, resignation is effective one year after notification, so Canada officially left WAHO in spring 1999. WAHO, however, approved Canada's reentry if breeders there wanted it and also stated that Canadian horses could be registered in the new American registry, the Purebred Arabian Horse Registry of America (PAHR), that had sprung up during the conflict to provide American pedigrees acceptable to WAHO.[36] Canadian breeders, however, knew where their welfare lay. The AHRA refused to allow the showing of Arabians that did not have an AHRA pedigree, and it did not recognize PAHR. The Canadian Arabian Horse Registry reaffirmed its allegiance to AHRA in the spring of 1998 and stated that it would not accept PAHR pedigrees either. The president of the Canadian association noted that the AHRA and CAHR had "a long history of cooperation and respect for one another."[37]

Other members of WAHO believed it made no sense to ostracize the largest single registry in the world or to threaten the world's largest market for Arabians. One country not eager to break off contact with the American registry was Jordan. Princess Alia of Jordan, president of the Royal Jordanian Stud, wrote to the executive committee of WAHO in 1997:

> The matter of AHRA being (potentially) suspended was sprung on us all in Abu Dhabi, and as was perfectly clear from the statements from the floor, it was not a popular one. We were not given the chance to really do anything about the matter, then or since. Terminating the AHRA international standing is simply not within a few individuals' power. I feel that for member countries to allow other registries to be replaced by the minions for an undemocratically elected committee is simply neither safe nor acceptable. As for justifying this ludicrous behavior by more or less saying that WAHO was turning a hypocritical blind eye by calling our horses purebred when most of them aren't and never were, is simply destructive and counter productive. It is also extremely insensitive and insulting to countries and registries who, through belief in our original common goals neglected and lost lines of genuine rare pure bred horses which are irreplaceable because of WAHO dictates.

"If WAHO is to continue," she warned, "I strongly recommend that it do so on rational calm and constructive lines. If not, then I know that I am speaking for many other registries worldwide, in saying that, I am not at all sure that we can continue to support it."[38] In 1998 Jordan and the AHRA signed a bilateral agreement for trade in purebred stock between the two countries.[39]

Individual breeders in the United States responded to the conflict in various ways. An open forum put on the Internet in 1998 revealed some opinions. One, for example, stated: "Is this truly in the name of caring about the breed, or is it merely a combination of self-promotion and the wish to pacify certain monied individuals?" Another wrote: "I for one do not understand the logic of WAHO's position. Once impure blood is introduced, the following offspring can never be 'purebred' unless they are trying to establish a 'new breed.' ... I agree with AHRA's viewpoint." It was apparent that some saw the conflict as a battle of interests that seemed to have little to do with the concerns of breeders. Most breeders, however, also seemed to support the idea of purity held by the AHRA and to wonder what WAHO was attempting to do.[40] To these breeders, purity, or the implication of it in pedigrees, remained fundamental to the idea of purebred breeding. The meshing of Eastern and Western views also permeated this point of view.

At the time I wrote this book, Jay Stream was still president of WAHO, and WAHO had not reinstated the AHRA. The world organization had managed to survive and to maintain a large membership. Recording of Arabians in the United States still took place chiefly through the AHRA. American pedigrees could find acceptance by WAHO, though, through registration with the American-based International Arabian Horse Organization, founded in 1950 through an AHRA motion. The AHRA believed that a separate organization was needed to promote and coordinate shows and to establish rules for showing. Although PAHR still existed, it lacked the strength of the International Arabian Horse Organization in American Arabian horse affairs, partially because the AHRA did not recognize it. The International Arabian Horse Organization, then, had been always been closely connected to the AHRA and had historically regulated shows. The WAHO issue drew it into registry matters. It maintained its close working ties with the AHRA, however, in spite of the dual registry system.[41]

By late 2001 the AHRA had succeeded in establishing a new Arabian horse breeding union. The registry completed negotiations with countries in the Americas—most of them also members of WAHO—to create an organization called the Alliance of the Americas. The alliance brought together Western Hemisphere breeders from Canada, the United States, Mexico, Brazil, Argentina, Paraguay, Uruguay, Colombia, Chile, Belize, Bolivia, Costa Rica, Cuba, Ecuador, Guatemala, Haiti, and Venezuela. The organization represented 75 percent of the world's Arabians. The AHRA now belonged to a large Arabian horse alliance, strengthening its position against WAHO. When it abandoned a policy of isolation, the AHRA also changed its earlier position on an issue that had precipitated the break with WAHO. The registry now granted pedigree status to certain South American horses previously designated as impure for recording purposes. "The Registry agreed, in forming this association, to provide registration documentation for certain bloodlines that it did not accept previously," the AHRA announced on 6 October 2001. The AHRA's new stance bore a surprising resemblance to WAHO's position on the status of Kurdo III's descendants, which traced their background to a GSB-registered horse in Kurdo III's pedigree. "These horses are considered to be purebred Arabians by nearly every other registry and in every country in the world, including the Arab countries," the AHRA stated. "They were imported to South America in good faith as purebred Arabians some 100 years ago[;] in the intervening time, they have been bred as purebred Arabians in South America and elsewhere." The registry now argued that purity had never been the main issue in the debate with WAHO. It had been, and still was, a question of sovereignty. The AHRA explained: "The Registry's expulsion from WAHO was unfortunate and ill-considered. The decision was taken unilaterally by WAHO due to the Registry's refusal to accept WAHO directives to accept all horses deemed by WAHO to be purebred. The WAHO dictate remains unacceptable to the Registry." It added, "The Registry was expelled from WAHO because the Registry refused to surrender its sovereign authority to make decisions about which horse and studbooks the Registry will accept."[42] The situation in the United States changed again early in 2003, at the time this book went to press. The AHRA and the International Arabian Horse Association joined to form a new organization called the Arabian Horse Association. The larger issue of pedigree control and WAHO, however, remains unresolved.[43]

The story of the WAHO-AHRA debate reveals a number of interesting patterns. First, it shows that in the world of purebred breeding, issues of purity and quality, and their meaning in pedigrees, are very much alive at the turn of the twenty-first century. They remain just as difficult to define or regulate. And they also remain intimately attached to market function. Second, the debate reveals that contentious issues about the role of wealth and self-interest in the purebred industry still concern breeders. But perhaps most important of all, the story of the debate demonstrates that attempts at world organization to overcome the problem of clashing national standards did not necessarily make the market work more effectively, nor could global association define what either quality or purity meant.

Record keeping in itself implies that the past can, and also should, govern the present and the future. The world's first public studbook, the GSB, sheltered the registering of Arabians in the nineteenth century by maintaining what would become a foundation public registry for the world's Arabians, with standards accepted anywhere. Over the first half of the twentieth century, record keeping for Arabians became separate from that for Thoroughbreds. Ironically, however, the old union would come back to complicate pedigree status for certain Arabians, affecting the interests of breeders all over the world. The accuracy of GSB records created the conflict between the AHRA and WAHO, and through that conflict it disrupted general relations among breeders in various countries—breeders who had no direct interest in the particular issue itself. Purity had once been guaranteed by the GSB. Now the GSB could be used to prove lack of purity. The concept of purity itself, however, remained fundamental to the breeding of Arabians. The horse was perceived to be pure when it left Arabia, and Western breeding from the beginning devoted itself to preserving that purity. Whether Western breeding could or should attempt improvement, in light of that purity, remains controversial, based on the link between quality and purity. If purity can be preserved, can improvement take place in breeding of the horses? And should it? How are these issues affected by the marketplace? These are questions faced by all breeders using the purebred system discussed in this book at all times.

# CONCLUDING REMARKS

Do purebred breeders produce animals more for the joy of breeding than for the money they generate? How can we answer this overarching question? Patterns in Shorthorn, Collie, and Arabian horse production over the past two hundred years show that market issues and breeding ideology were and are irrevocably intertwined in modern purebred breeding. Absolute ideology cannot be detached from absolute market concerns, and to comprehend the dynamics of this breeding system one must appreciate that fact. This book demonstrates that the market and breeding matters that drove modern purebred breeding resulted from a specific relationship between two countries: Britain and the United States. While tangled market and breeding problems explain characteristics inherent in the system, the industry's fundamental structures grew in response to the demands created by a trade connection between Britain and the United States.

Breeders were aware of the dichotomy that the modern purebred industry presented: breeding for financial gain or to attain an ideal. Sometimes they tried to separate the two issues. For example, they argued (correctly) that the ability to spend vast amounts of money on good animals did not make a person a good breeder. But while that might be true, it did not prove that market demands were unrelated to breeding practices. A Collie breeder said: "It is easy with an unlimited bank account to assemble a group of winning dogs of whatever breed. It is a long and hard and serious task to mate those dogs judiciously, to rear and develop their progeny to bring them to the bloom of their greatest potential excellence, and so to go on from generation to generation, discarding the inferior and retaining the superior in the never-ending quest for perfection."[1] The process of producing good stock might itself have fascinated breeders in the way he claimed, but money clearly remained connected to the desire to excel. Selling played a role in the quest for "perfection." The market confirmed the level of acceptance that a breeder's judgment of quality could command and therefore reflected the degree of "perfection" achieved. The market was critical, not just from a monetary point of view, but from a breeding perspective as well.

Breeders worried about the entanglement of breeding ideology with market forces. The utility/fancy conflict arose out of this problem and forced breeders to assess the relation of trade factors to the production of better quality. Did show competition and its contingent marketing effects enhance or undermine the basic value of original type, usually related to the animal's usefulness? How did monetary value affect breeding decisions with respect to these problems? Could too much of a good thing—a healthy international market—really be a bad thing? Did the desire for Duchess cattle in the 1860s merely reflect investment interests, or was it shaped by love of the beautiful cattle that Bates had bred? Did Morgan's buying change the shape of the Collie's head and increase the dog's viciousness? Did the high-priced auctions of Arabians in the 1980s lead to the production of very ordinary horses through the pursuit of new or exotic genetics fueled from abroad? Was purebred breeding about an ideal or about selling a commodity? Such questions concerned breeders, and we cannot see how entangled the issues became in people's minds without addressing what breeders tried to achieve through actual animals.

That buying and selling played such an essential part in modern purebred breeding meant that booms could emerge. When individuals became adept at taking advantage of inflated markets, there were serious ramifications. Opportunities for sales in an overheated market allowed breeders with good marketing skills to capitalize by selling for high sums a product they had spent years perfecting. That environment in turn would stimulate yet another boom, and new players would appear to trade in the commodity. These people, however, were not true breeders. They were better defined as investors. It was often the lack of clear demarcation between breeder and investor that made market factors look as if they alone drove the purebred industry. Hot markets could mask underling issues that related to breeding alone. Booms did not in themselves explain what motivated people to breed. Booms explained why people wanted to invest.

The breeding system described in this book took its fundamental shape from the fusion of two elements. A desire in eighteenth-century Britain to standardize type in animals led to systematic inbreeding, with emphasis on selection through male lines. The method became connected with the recording of pedigrees in public studbooks. A transatlantic trade in agricultural animals showed how well this connection worked in the marketplace. Public records made trade over long distances easier. Agriculture played a major role in the expansion of the system, because farming drove a general market for purebred animals on an international scale. The keeping of public records would ultimately govern how international markets worked and how market forces influenced the way animals were bred. As other factors interacted with the system—perceptions about genetics and eugenics, attitudes toward art and ideal type, and conflicting views about the meaning of improvement, quality, and purity—studbooks played an even greater role in actual methods of breeding.

Over time, certain characteristics clung persistently to breeding practices. Breeders from the beginning concerned themselves with appearance. But faulty attributes that might be associated with desirable looks were generally ignored. Infertility in Duchess Shorthorns and savage temperament in Collies, for example, would be tolerated if the animals' appearance matched the ideal that commanded the attention of the market. Breeders positively avoided looks that suggested foreign blood. Since color often was taken to reveal that characteristic, it influenced breeders. The proper color designated high quality because it suggested purity—freedom from the blood of other breeds. Emphasis on breeding from male lines tended to hide the genetic input of females. Certain animals, generally males, established breeding lines that would be valued over generations. Identifiable ancestry behind these sires, however, often traced to females whose genes passed down through their sons. When Western breeding theory became linked with Eastern—as happened with the Arabian horse—the breeding system became more complicated. Emphasis on males evolved within a breed that had been traditionally structured on female lines. Families had also been emphasized in the production of this horse in Arabia, whereas Western thinking stressed the importance of individual animals in breeding programs. Arabian horse breeding culture would ultimately take an unclear stand between Eastern and Western methods. Individual stallions could often be revered at the same time that classic strain or family breeding was touted.

Different systems for pedigree recording evolved in various countries. While Canada's pedigree regulations remained sensitive to changes in either the United States or Britain, for example, the structure for purebred breeding that emerged there relied on some government supervision, which did not occur in Britain or the United States. Curiously, a somewhat similar pattern took root in other countries, affecting Arabian horses in Egypt after 1914 and in Poland and Russia after the First World War. The size of the market in these countries relative to the larger American market and international market may have led to this government support system.

While registry systems originated to help identify animals, information on hereditary background quickly came to influence breeding itself. Breeders became less aware that pedigrees played as much of a role in the buying and selling of stock as they did as breeding tools. Breeders believed that pedigrees should guarantee characteristics, yet it remained hard to define how, or even what those characteristics might be. Purity became critical to breeders, but it could be difficult to separate purity from quality. Was quality more important than purity, and how should one describe quality? What if quality diverged from the original type? Could divergence be defined as quality? Did type relate to authenticity and to use? Because so many of these questions related to the value and meaning of pedigrees, they could be answered in many ways, and answers changed as standards changed.

Too much allegiance to pedigrees as a guarantee of purity would lead to severe problems, as the Duchess boom made clear. The Duchess line became extinct through excessive concern with pedigree. The history of international Arabian horse breeding shows that obsession with pedigrees as a guarantee of quality could interfere with the way the market worked and even with the ability to define quality in Arabians. If breeders insisted on purity, and if records were considered evidence of purity yet could not prove that purity existed, the whole system might collapse. In the early days of breeding under the system, certain problems arose out of misunderstood genetic traits. But even as heredity has become better understood, it has not been easier to specify what purity or quality means. In fact genetics has done surprisingly little to influence the way this purebred breeding method works or to influence the animals' marketability.

A purebred breeding system began within the Anglo world and took its ultimate shape there as well, because of trade ties. The system spread to other parts of the world when the international market drew other countries into a trade that had evolved between Britain and the United States. As historical Arabian horse breeding shows, breeders who wished to join in international trade found it easier to do so if they adopted the concept of a public registry. Inbreeding and the practice of highlighting males in breeding would become culturally embedded in Arabian horse breeding where public recording had been established. Greater concern over what role inbreeding played in strain theory is one example of that cultural pattern. While Europeans traditionally used stallions to upgrade local horses (upgrading was intensely done in Poland), emphasis on males in breeding to define family lines is another example, because the pattern became increasingly common where public, rather than private, recording took place.

There have been very few outstanding breeders over the years, which is probably one reason fads have taken such a tenacious hold at various times. Fads also explain certain forms of intense inbreeding, emphasis on males in breeding, and concern with pedigrees. Inbreeding was not always done to achieve "improvement." Emphasis on males in breeding programs and the use of pedigrees in selection were not always practiced to "improve" the animals. Inbreeding preserved the breeding skills of lost masters, and pedigrees guaranteed that the master's hand lay behind the animal's makeup. Ideas about the purity of the Duchesses illustrate this: breeders attempted to save the purity of Bates's art. The compulsion to inbreed to stock descended from Abbas Pasha breeding is another example. Inbreeding also preserved authenticity, demonstrating efforts to maintain historical characteristics rather than to improve animals. The thinking of breeders who inbred to Skowronek demonstrated both these types of reasoning. The stallion represented perfection and authenticity to original type. Arabian breeders hoped to perpetuate the ideal, assuming that they could not improve on it. Out of a desire to preserve the past, certain horses were labeled as "blue star" to authenticate them as representatives of original breeding. Authenticity, purity, rarity, and verification of the presence of these traits have always been marketable commodities.

I have explained aspects of the development and practice of a breeding system through a study of the fortunes of certain breeds in certain countries. The countries and breeds discussed were not selected randomly; they were chosen because particular locales and breeds best illustrated underlying patterns owing to their centrality or to the time when critical developments took place. A more detailed study of how much breeding practices in other nations were influenced by the formalized inbreeding methods set under the Anglo system would be of considerable interest, as would the history of many other breeds. Although market forces and use of various components of the breeding system described here might combine to create new breeds or preserve older ones, this work does not attempt to explain the evolution of the thousands of other breeds or to assess their histories in relation to each other. Many more stories could be told. The interaction of breeds within and across countries, for example, would be an interesting topic to explore. Comparing how standards for pedigrees evolved in different species would be worthwhile. The structure of organizations that helped to regulate breeders also varied from species to species as well as from country to country, and not enough is known about them or about the influence of markets and regulation on the ebb and flow of popularity in various breeds. The interplay of science and culture in purebred breeding could be better understood as well, and a more focused perspective on the impact of either one would tell us more about factors influencing the system.

The purebred industry has supported a large and complicated market. But purebred breeding is and has always been motivated by the love of animals and the skill of breeding, a love that is rewarded by ability to sell the stock. The system still revolves around using science and art in pursuit of a vision of excellence, in order to survive in the marketplace. Breeders test the "perfection" of their work by assessing how salable the product is. Modern purebred breeding, then, is in many ways about the art of genetics and about the marketing of that art form.

# NOTES

## One. Developing a Modern Method of Purebred Breeding

1. J. Serpell, ed., *The Domestic Dog: Its Evolution, Behavior, and Interactions with People* (Cambridge: Cambridge University Press, 1995), 8, 10; T. Grandin, ed., *Genetics and the Behavior of Domestic Animals* (London: Academic Press, 1998), 206; R. Coppinger and L. Coppinger, *Dogs* (New York: Scribner, 2001), 316.

2. N. Russell, *Like Engend'ring Like: Heredity and Animal Breeding in Early Modern England* (Cambridge: Cambridge University Press, 1986), 2–4.

3. Serpell, ed., *Domestic Dog*, 15, 33, 37, 38; D. Goodall, *A History of Horse Breeding* (London: Robert Hall, 1977), 33; Grandin, *Genetics and the Behavior of Domestic Animals*, 20–38, 35, 48; Coppinger and Coppinger, *Dogs*, 54–55, 66, 279, 312–14. The Coppingers argue that dogs domesticated themselves; see 39–67.

4. J. Clutton-Brock, *A Natural History of Domesticated Animals*, 2d ed. (Cambridge: Cambridge University Press, 1999), 40.

5. Clutton-Brock, *Natural History*, 47.

6. K. Thomas, *Man and the Natural World: Changing Attitudes in England, 1500–1800* (London: Allen Lane, 1983), 192.

7. Thomas, *Man and the Natural World*, 223.

8. For eighteenth-century crop and animal husbandry, see G. Fussell, "Animal Husbandry in Eighteenth-Century England," *Agricultural History Review* (hereafter *AHR*) 11 (1937): 96–116, and G. Fussell, "Crop Husbandry in Eighteenth-Century England," *AHR* 15 (1941): 202–16, and 16 (1942): 41–63.

9. For hybridizing as the means of all changes in domesticated plants and animals, see Coppinger and Coppinger, *Dogs*, 316. For degeneration, see W. E. Castle, "Biological and Social Consequences of Race-Crossing," *Journal of Heredity* 15 (1924): 363–69.

10. For a full discussion of hybridizing and the breeding of animals, see H. Ritvo, *The Platypus and the Mermaid and Other Figments of the Classifying Imagination* (Cambridge: Harvard University Press, 1997), 85–130.

11. O. Lloyd-Jones, "What Is a Breed?" *Journal of Heredity* 6 (1915): 531–37.

12. See, for example, attempts by Border Collie breeders to define "breed" in "How Many Breeds?" *United States Border Collie Club Newsletter*, Summer 1995, and "What Is a Breed, Anyway?" *United States Border Collie Club Newsletter*, Spring 1995.

13. J. Walton, "Pedigree and the National Cattle Herd circa 1750–1950," *AHR* 34 (1986): 152.

14. H. Ritvo, *The Animal Estate* (Cambridge: Harvard University Press, 1987), 66–67; H. C. Pawson, *Robert Bakewell, Pioneer Livestock Breeder* (London: Crosby Lockwood, 1957), 51–52, 61–62, 68, 90–91, 93; R. Trow-Smith, *A History of British Livestock Husbandry, 1700–1900* (London: Routledge and Kegan Paul, 1959), 49, 51–54.

15. Trow-Smith, *History of British Livestock Husbandry*, 54.

16. Goodall, *Horse Breeding*, 218.

17. B. Tozer, *The Horse in History* (London: Methuen, 1908), 203–4.

18. J. Clutton-Brock, *Horse Power: Horse and Donkey in Human Societies* (London: Natural History Museum Publications, 1992), 61.

19. R. Archer, C. Pearson, and C. Covey, *The Crabbet Arabian Stud: Its History and Influence* (Northleach, U.K.: Alexander Heriot, 1978), 35.

20. *The General Stud-Book, Containing Pedigrees of Race Horses, from the Earliest Accounts to the Year 1807, Inclusive* (London: Printed for James Weatherby by H. Reynell, 1808), 516.

21. Russell, *Like Engend'ring Like*, 19, 95–96; W. Vamplew, *The Turf* (London: Allen Lane, 1976), 78–80; R. Longrigg, *The History of Horse Racing* (London: Macmillan, 1972), 91–92; M. Huggins, *Flat Racing and British Society, 1790–1914* (London: Frank Cass, 2000); *General Stud Book*, vol. 1.

22. J. L. Lush, *Animal Breeding Plans* (Ames, Iowa: Collegiate Press, 1937), 24.

23. W. Houseman, *The Improved Shorthorn: Notes and Reflections upon Some Facts in Shorthorn History, with Remarks upon Certain Principles of Breeding* (London: Ridgeway, 1876), 20.

24. Lush, *Animal Breeding Plans*, 24.

25. E. Whetham, "The Trade in Pedigree Livestock, 1850–1910," *AHR* 27 (1979): 47–50; A. Fraser, *Animal Husbandry Heresies* (London: Crosby Lockwood, 1960), 35; M. Lerner and H. Donald, *Modern Developments in Animal Breeding* (New York: Academic Press, 1966), 156; Lush, *Animal Breeding Plans*, 26; E. Whetham, *The Agrarian History of England and Wales, 1914–38*, vol. 8 (Cambridge: Cambridge University Press, 1978), 7.

26. O. Lloyd-Jones, "What Is a Breed?" *American Breeders' Magazine* 6 (1915): 536–37; G. MacEwan, *Heavy Horses: Highlights of Their History* (Saskatoon, Sask.: Western Producer Prairie Books, 1986), 42–43, 50–52.

27. Lady Wentworth, *Thoroughbred Racing Stock* (New York: Charles Scribner's Sons, 1938), 60.

28. Ontario, Legislature, Sessional Paper (hereafter SP, Ontario), 13, 1875–76, 31–32; SP 12, Ontario, 1877, 48.

29. *Farmer's Advocate*, January 1876, 13.

30. *Farming*, February 1896, 337.

31. Ritvo, *Animal Estate*, 302.

32. See P. Thurtle, "The Creation of Genetic Identity," electronic version

of the *Stanford Humanities Review,* vol. 5, supplement, updated 17 December 1996, www.stanford.edu/group/SHR/shreview/5-supp/text/thurtle.html, accessed 29 August 1999. See also B. Kimmelman, "The American Breeders' Association: Genetics and Eugenics in an Agricultural Context, 1903–1913," *Social Studies of Science* 13 (1983): 163–204.

33. *American Breeders' Magazine* 1 (1910): 65.

34. *American Breeders' Magazine* 1 (1910): 65.

35. *Farming World,* 15 August 1907, 748.

36. *American Breeders' Magazine* 3 (1912): 271.

37. E. B. Babcock and R. E. Clausen, *Genetics in Relation to Agriculture* (New York: McGraw-Hill, 1918), 12.

38. D. F. Jones, *Genetics in Plant and Animal Improvement* (New York: John Wiley, 1925), 486, 487.

39. A. J. Clarke, ed., *Animal Breeding: Technology for the 21st Century* (Amsterdam: Harwood, 1998), 7.

40. A. H. Thompson, "Starbuck Reborn: The Clone of the Famed Canadian Bull Hanover Hill Starbuck Is Now Two Years Old," *Ontario Dairy Farmer* 16 (October 2002): 68.

41. P. Mazumdar, *Eugenics, Human Genetics and Human Failings: The Eugenics Society, Its Sources and Critics in Britain* (London: Routledge, 1992), 3, 15, 58–59.

42. Thurtle, "Creation of Genetic Identity."

43. Mazumdar, *Eugenics, Human Genetics and Human Failings,* 58–59, 71.

44. A. McLaren, *Our Own Master Race: Eugenics in Canada, 1885–1945* (Toronto: McClelland and Stewart, 1990), 16, 17.

45. M. Derry, *Ontario's Cattle Kingdom: Purebred Breeders and Their World, 1870–1920* (Toronto: University of Toronto Press, 2001), 24–25.

46. Jones, *Genetics in Plant and Animal Improvement,* 486, 487.

47. Housman, *Improved Shorthorn,* 9.

48. *Farmer's Advocate,* July 1880, 164.

49. Derry, *Ontario's Cattle Kingdom,* 25–39.

50. W. C. Hill and T. F. C. Mackay, eds., *Evolution and Animal Breeding: Reviews on Molecular and Quantitative Approaches in Honour of Alan Robertson* (Wallingford, U.K.: C. A. B. International, 1989), 3–5, 155–59, 237, 241.

## *Two.* Shorthorns and Animal Improvement

1. A. Sanders, *Short-Horn Cattle: A Series of Historical Sketches, Memoirs and Records of the Breed and Its Development in the United States and Canada* (Chicago: Sanders, 1900), 14.

2. See D. C. Moore, "The Corn Laws and High Farming," *Economic History Review* (hereafter *EHR*), 2d ser., 18 (1965): 544–61.

3. T. W. Fletcher, "The Great Depression of English Agriculture, 1873–1896," *EHR,* 2d ser., 13 (1961): 431.

4. R. Trow-Smith, *A History of British Livestock Husbandry, 1700−1900* (London: Routledge and Kegan Paul, 1959), 57, 84, 89−116.

5. Sanders, *Short-Horn Cattle*, 31, 34−35, 37−39, 44; see S. Wright, "Mendelian Analysis of the Pure Bred Breeds of Livestock, Part 2, The Duchess Family of Shorthorns as Bred by Thomas Bates," *Journal of Heredity* 14 (1923): 405.

6. Sanders, *Short-Horn Cattle*, 75.

7. See Wright, "Mendelian Analysis," 405−22.

8. Sanders, *Short-Horn Cattle*, 81−85.

9. J. Lush, "Notes on Animal Breeding," manuscript, 1933, University of Guelph, chap. 3, 61.

10. H. Ritvo, *The Animal Estate* (Cambridge: Harvard University Press, 1987), 61.

11. *Proceedings of the Fourth Annual Convention of the American Association of Breeders of Short Horns, Toronto, 1875* (Toronto: Globe, 1875), 22.

12. For more on Spencer and agriculture, see E. A. Wasson, "The Third Earl of Spencer and Agriculture, 1818−1845," *Agricultural History Review* (hereafter *AHR*) 26 (1978): 89−99.

13. J. Walton, "Pedigree and the National Herd circa 1750−1950," *AHR* 34 (1986): 149−70. See also P. J. Perry, "The Shorthorn Comes of Age, 1822−1843," *Agricultural History* (hereafter *AH*) 56 (1982): 560−66.

14. P. Henlein, "Cattle Driving from the Ohio Country, 1800−1850," *AH* 28 (1954): 94; "Shifting Range-Feeding Patterns in the Ohio Valley before 1860," *AH* 31 (1957): 1−11; J. Whitaker, *Feedlot Empire: Beef Cattle Feeding in Illinois and Iowa, 1840−1900* (Ames: Iowa State University Press, 1975), 55, 64, 82.

15. C. B. Plumb, "Felix Renick, Pioneer," *Ohio Archaeological and Historical Publications* 38 (1924): 21−22, 28−30, 35−41.

16. P. Henlein, "Cattle Kingdom in the Ohio Valley: The Beef Cattle Industry in the Ohio Valley, 1783−1860" (Ph.D. diss., University of Wisconsin, 1957), 93.

17. Sanders, *Short-Horn Cattle*, 199−202.

18. Sanders, *Short-Horn Cattle*, 214.

19. Plumb, "Felix Renick, Pioneer," 35−50.

20. Sanders, *Short-Horn Cattle*, 263.

21. G. MacEwan, *Highlights of Shorthorn History* (Winnipeg, Man.: Hignill, 1982), 36, 70.

22. Sanders, *Short-Horn Cattle*, 263, 379−87, 393.

23. M. Derry, *Ontario's Cattle Kingdom: Purebred Breeders and Their World, 1870−1920* (Toronto: University of Toronto Press, 2001), 5−6.

24. R. L. Jones, *Agriculture in Ohio to 1880* (Kent, Ohio: Kent State University Press, 1983), 116; F. Shannon, *The Farmer's Last Stand: Agriculture, 1860−1897* (New York: Holt, Rinehart and Winston, 1961), 199−200.

25. Canada, Parliament, Sessional Paper (hereafter SP, Canada), 10, 1913, 548.

26. Derry, *Ontario's Cattle Kingdom*, 15—16, 18—19.

27. D. Marshall, *Shorthorn Cattle in Canada* ([Toronto]: Dominion Shorthorn Breeders' Association, 1932), 117—26, 211.

28. *Proceedings . . . American Shorthorn Breeders' Association* 1876, 95.

29. Sanders, *Short-Horn Cattle*, 450.

30. Sanders, *Short-Horn Cattle*, 510, 712—13, 720.

31. W. Housman, *The Improved Shorthorn Notes and Reflections upon Some Facts in Shorthorn History, with Remarks upon Certain Principles of Breeding* (London: Ridgeway, 1876), 31, 36—37, 3.

32. *Proceedings . . . American Shorthorn Breeders' Association*, 1876, 48.

33. *Farmer's Advocate*, February 1876, 27.

34. Ontario, Legislature, Sessional Paper (hereafter SP, Ontario), 26, Ontario, 1897, 127.

35. A. Sanders, *Red, White and Roan* (Chicago: American Shorthorn Breeders' Association, 1936), 33.

36. Sanders, *Short-Horn Cattle*, 846.

37. I. M. Bruce, *The History of the Aberdeenshire Shorthorn* (Aberdeen: Aberdeen, Banff and Kincardine Shorthorn Breeders' Association, 1923), 100—104.

38. Cruickshank to Davidson, 21 May 1876, Miller-Davidson Papers, Archives, University of Guelph.

39. Bruce, *Aberdeenshire Shorthorn*, 603.

40. Cruickshank to Davidson, 21 May 1876, Miller-Davidson Papers.

41. Cruickshank to Davidson, 25 September 1888, Miller-Davidson Papers.

42. Cruickshank to Davidson, 8 January 1882, Miller-Davidson Papers.

43. Cruickshank to Davidson, 17 June 1887, Miller-Davidson Papers.

44. *Canadian Live-Stock and Farm Journal*, September 1885, 226; July 1886, 176.

45. *Canadian Live-Stock and Farm Journal*, March 1885, 59.

46. Arthur Johnston to a Quebec breeder, 27 May 1902, Arthur Johnston Papers, Letterbook 7, Ontario Archives.

47. SP 20, Ontario, 1895, 47.

48. Sanders, *Short-Horn Cattle*, 110, 114.

49. Great Britain, *Journal of the Ministry of Agriculture* 28 (May 1921): 110—11.

50. J. Lush, *Animal Breeding Plans* (Ames, Iowa: Collegiate Press, 1937), 26; A. Fraser, *Animal Husbandry Heresies* (London: Crosby Lockwood, 1960), 35; M. Lerner and H. Donald, *Modern Developments in Animal Breeding* (New York: Academic Press, 1966), 156; E. Whetham, "The Trade in Pedigree Livestock, 1850—1910," *AHR* 27 (1979): 47—50.

51. Ritvo, *Animal Estate*, 60—61.

52. *Canadian Live-Stock and Farm Journal*, January 1887, 368.

53. Sanders, *Short-Horn Cattle*, 282—83.

54. Lewis Allan, Circular of the American Short-Horn Herd Book, Buffalo, N.Y., October 1874, Canadian Institute for Historical Microreproductions (CIHM 53873).

55. SP 28, Ontario, 1896, 174.

56. Sinclair, *History of Shorthorn Cattle* (London: Vinton, 1907), 87.

57. Sinclair, *History of Shorthorn Cattle*, 673.

58. SP 3, Ontario, 1882–83, appendix B, 151–52.

59. *Farmer's Advocate*, 11 March 1909, 380–81.

60. SP 10, Canada, 1880, 119, 135, 138.

61. *Report of the Ontario Agricultural Commission* 4 (1880): 6, 11, 19.

62. SP 28, Ontario, 1896, "History of the Agriculture and Arts Association," 164, 174–75.

63. *Canadian Live-Stock and Farm Journal*, February 1885, 31.

64. *Farmer's Advocate*, June 1880, 135.

65. *Farmer's Advocate*, June 1880, 135; March 1881, 65; July 1881, 157.

66. *Farmer's Advocate*, July 1881, 157; October 1881, 248; December 1881, 294.

67. *Canadian Live-Stock and Farm Journal*, April 1887, 459.

68. *Canadian Live-Stock and Farm Journal*, February 1887, 31.

69. *Farmer's Advocate*, April 1886, 105–6.

70. Looking at essays on prize farms, which were published in the Ontario Legislature's Sessional Papers in the 1890s, makes this statement abundantly clear.

71. *Farmer's Advocate*, July 1892.

72. *Canadian Live-Stock and Farm Journal*, July 1895, 147.

73. Arthur Johnson to Fisher, 19 January 1901, Arthur Johnson Papers, Letterbook 5.

74. *Collie Folio*, February 1911, 62.

75. *Farming World and Canadian Farm and Home*, 15 April 1905, 293.

76. *Farming World*, 17 December 1901, 699.

77. United States Department of Agriculture, *Report*, 1906, 29, 156; 1908, 266–67; *Field and Fancy*, 4 August 1906, 4.

78. See SP 28, Ontario, 1898–99, 124, 126.

79. SP 24, Ontario, 1905, 19.

80. SP 73, Ontario, 1900, 16.

81. SP 24, Ontario, 1901, 5–6, 8, 12.

82. SP 24, Ontario, 1902, 10.

83. *Collie Folio*, February 1911, 61.

84. *Yearbook of the Department of Agriculture* (United States), 1913, 514.

85. United States Department of Agriculture, *Report*, 1905, 55.

86. United States Department of Agriculture, *Report*, 1908, 267.

87. *Statutes*, 1900, chap. 33, "An Act Respecting the Incorporation of Live Stock Records Associations."

88. SP 24, Ontario, 1905, 12, 18, 44, 52.

89. *Farming World and Canadian Farm and Home*, 15 April 1905, 293, 1 April 1904, 251.

90. SP 15, Canada, 1906, xxxix.

91. *Farmer's Advocate*, 11 April 1912, 684.

92. *Farming World and Canadian Farm and Home*, 15 March 1905, 213.

93. *Wallace Farmer* quoted in *Farming World and Canadian Farm and Home*, 15 June 1905, 454.

94. *Farmer's Advocate*, November 1886, 334; Johnston to F. W. Hodson, June 24 1896, Arthur Johnston Papers, Letterbook 1.

95. "Minutes of Evidence Taken before the Department Committee Appointed to Enquire and Report as to the British Trade in Live Stock with the Colonies and Other Countries," 1912, Cd. 6032 (Britain), 14.

96. *Yearbook of the Department of Agriculture* (United States), 1919, 423.

97. *Yearbook of the Department of Agriculture* (United States), 1919, 408, 409, 411, 414, 418, 421; 1914, 100.

98. *Agricultural Gazette* 5 (1918): 300.

## *Three.* Producing Beautiful Dogs

1. *Kennel Gazette* (British), April 1880, 3.

2. Secord, *Dog Painting, 1840—1940: A Social History of the Dog in Art* (Woodbridge, U.K.: Antique Collectors' Club, 1992), 51—55.

3. *Kennel Gazette* (British), November 1883, 489.

4. "Greyhound History," Greyhound Pets in America, www.fastfriends .org/the-grey-list.html, accessed 8 October 2000.

5. K. White, "Victorian and Edwardian Dogs," *Veterinary History*, n.s., 7 (1992): 72—77.

6. *Illustrated London News*, 8 February 1851, 99.

7. *Illustrated London News*, 8 February 1851, 100.

8. E. Montcrieff with S. Joseph and I. Joseph, *Farm Animal Portraits* (Woodbridge, U.K.: Antique Collectors' Club, 1996), 280—83, 285—87.

9. Montcrieff, *Farm Animal Portraits*, 25.

10. G. Burnham, *The History of the Hen Fever* (1855; reprint, San Diego: Frank E. Marcy, 1935), 24, 25.

11. *Illustrated London News*, 22 January 1853, 59.

12. Burnham, *Hen Fever*, 102—3, 106—7; Moncrieff, *Farm Animal Portraits*, 285.

13. *Farmer's Advocate*, February 1869, 25.

14. *Canadian Poultry Review*, March 1878, 51.

15. *Farming*, 11 January 1898, 148; 12 December 1899, 408; *Farmer's Advocate*, 25 May 1905, 782; *Canadian Live-Stock Journal*, June 1895, 136; *Canadian Poultry Review*, April 1878, 69; March 1879, 55; November 1878, 179.

16. *Farmer's Advocate*, 20 April 1905, 585.

17. *Farmer's Advocate*, 15 August 1900, 474—75; 12 August 1915, 1275; *Canadian Poultry Review*, November 1878, 179; "Minutes of Evidence," 45, 48, 105; SP 39, Ontario, 1916, 39.

18. M. G. Denlinger, *The Complete Collie*, 3d ed. (Richmond, Va.: Denlinger's, 1949), 20.

19. *Poultry Chronicle*, March 1872, 136.

20. *Canadian Breeder and Agricultural Review*, 14 August 1885, 521.

21. See, for example, *Canadian Poultry Review*, December 1877, 15; May 1878, 103; December 1878, 9; January 1885; *Kennel Gazette* (Canadian), February 1889, 3.

22. See, for example, *Canadian Poultry Chronicle*, August 1870, 32.

23. I. Combe, D. Bridge, and P. Hutchinson, *Rough Collies of Distinction: A Pictorial Record of Influential British Rough Collies from the 1860s to the Present Day* (London: I. Combe, D. Bridge, and P. Hutchinson, 2001), 28.

24. G. Kaye, "A Century of Collies and Significant Developments: Part 1," *Collie Expressions*, January 2000, 31.

25. K. Marshall, *His Dogs: Albert Payson Terhune and the Sunnybank Collies* (New York: Collie Club of America Foundation [Collie Health Foundation], 2001), 46.

26. F. Jackson, *Crufts: The Official History* (London: Pelham Books, 1990), 39–42, 48.

27. M. E. Thurston, *The Lost History of the Canine Race* (New York: Avon Books, 1996), 110–11.

28. W. Secord, *Dog Painting: The European Breeds* (Woodbridge, U.K.: Antique Collectors' Club, 2000), 26, 34, 37.

29. Kennel Club, *Treasures of the Kennel Club: Paintings, Personalities, Pedigrees and Pets* (London: Kennel Club, 2000), 55–56.

30. Kennel Club, *Treasures of the Kennel Club*, 24, 69, 76.

31. R. Leighton, *The New Dog Book: British Dogs and Their Foreign Relatives, with Chapters on Law, Breeding, Kennel Management, and Veterinary Treatment* (London: Cassell, 1907), 643.

32. W. E. Jacquet, *The Kennel Club: A History and Record of Its Work* (London: Kennel Gazette, 1905), 8.

33. *Kennel Gazette* (British), April 1880, 3.

34. *Kennel Gazette* (British), August 1881, 340.

35. *Kennel Gazette* (British), June 1883, 373.

36. C. A. T. O'Neill, ed., *The American Kennel Club* (New York: Howell Book House, 1984), 10, 13–14, 16.

37. *Canadian Breeder and Agricultural Review*, 22 May 1885, 329.

38. *Kennel Gazette* (Canadian), February 1889, 6.

39. *Kennel Gazette* (Canadian), October 1889, 73; November 1889, 87.

40. *Field and Fancy*, 4 August 1906, 14.

41. *Collie Folio*, February 1911, 62.

42. *Agricultural Gazette* 2 (1915): 908.

43. *Collie Folio*, April 1914, 111.

44. *Agricultural Gazette* 2 (1915): 909; 5 (1918): 186.

45. *Kennel Gazette* (Canadian), August 1889, 50, 54.

46. O. P Bennett and T. M. Halpin, *The Collie* (Washington, Ill.: O. P. Bennett, 1942), 164—65.

47. *Kennel Gazette* (British), August 1888, 173.

48. H. Dalziel, *British Dogs: Describing the History, Characteristics, Points, and Club Standards, of the Various Breeds of Dogs Established in Great Britain,* 2d ed., vol. 2 (London: L. Upcott Gill, 1889), 24.

49. Dalziel, *British Dogs,* 26.

50. See also H. Ritvo, *The Animal Estate* (Cambridge: Harvard University Press, 1987), 83—115, and her article "Pride and Pedigree: The Evolution of the Victorian Dog Fancy," *Victorian Studies* 29 (1986): 227—53, for a discussion on dog breeding and shows.

51. J. Blunt Lytton (Lady Wentworth), *Toy Dogs and Their Ancestors* (London: Duckworth, 1911), 87, 112, 113, 279—81.

52. *Collie Folio,* March 1908, 83.

53. *Collie Folio,* March 1908, 84.

54. E. Humphrey and L. Warner, *Working Dogs: An Attempt to Produce a Strain of German Shepherds Which Combines Working Ability and Beauty of Conformation* (Baltimore: Johns Hopkins Press, 1934), 28.

55. Humphrey and Warner, *Working Dogs,* 40.

56. See R. Coppinger and L. Coppinger, *Dogs* (New York: Scribner, 2001).

57. Leighton, *New Dog Book,* 545—50.

58. *Ladies' Kennel Journal* 1 (December 1894): 1.

59. *Ladies' Kennel Journal* 5 (August 1897): 357.

60. See, for example, *Canadian Breeder and Agricultural Review,* 22 May 1885, 329; 6 August 1885, 505; and *Canadian Poultry Review,* April 1886, 86.

## *Four.* Patterns in Collie Breeding and Culture

1. H. Dalziel, *The Collie: As a Show Dog, Companion, and Worker,* 4th ed., rev. J. Maxtee (London: Bazaar, Exchange and Mart Office, 1921), 5—6.

2. See, for example, R. Cooper, *Catalogue of "Scotch Collies," Imported and Selected by R. Cooper, from the Best Breeders in Scotland* (Reading, Ohio, 1881).

3. M. DeVine, *Border Collies* (New York: Barron's Educational Series, 1997), 7—8.

4. DeVine, *Border Collies,* 11.

5. D. C. Coile, *Australian Shepherds* (New York: Barron's Educational Series, 1999), 6—8, 105—8.

6. H. Dalziel, *The Collie: Its History, Points and Breeding* (London: L. Upcott Gill, 1888), 1; Dalziel, *Collie: As a Show Dog,* 2.

7. Dalziel, *Collie: Its History,* 1; Dalziel, *Collie: As a Show Dog,* 2.

8. Dalziel, *British Dogs: Describing the History, Characteristics, Points, and Club Standards of Various Breeds of Dogs Established in Great Britain*, 2d ed. (London: L. Upcott Gill, 1889), 2:26.

9. M. M. Palmer, "The Scotch Collie, the Dog of Yesterday, To-day and To-morrow," *Country Life in America*, 1 July 1903, n.p.

10. R. Coppinger and L. Coppinger, *Dogs* (New York: Scribner, 2001), 57−58, 1−7, 71, 275, 279, 282, 289, 293−95, 311−16.

11. J. Watson, *The Dog Book: A Popular History of the Dog, with Practical Information as to Care and Management of House, Kennel, and Exhibition Dogs; and Descriptions of All the Important Breeds* (Toronto: Musson, 1906), 348; G. Kaye, "Collie Origins and History," www.chelsea-collies.com/History.html, accessed 25 July 1999; M. G. Denlinger, *The Complete Collie* (New York: Howell Book House, 1962), 15, 17; A. K. Nicholas, *The Collie* (Neptune City, N.J.: T. F. H. Publications, 1986), 12.

12. *Collie Folio*, August 1909, 277−78.

13. O. P. Bennett and T. M. Halpin, *The Collie* (Washington, Ill.: O. P. Bennett, 1942), 26.

14. *Collie Folio*, March 1910, 84.

15. Bennett, *Collie*, 142−43.

16. *Canadian Poultry Review*, December 1878, 9−10.

17. *Canadian Breeder and Agricultural Review*, 14 August 1885, 521.

18. Bennett, *Collie*, 142−43.

19. Bennett, *Collie*, 165.

20. *Collie Folio*, August 1910, 270.

21. Bennett, *Collie*, 21.

22. *Farmer's Advocate*, 16 January 1899, 36.

23. J. Lytton, *Toy Dogs and Their Ancestors* (London: Duckworth, 1911), 298.

24. Bennett, *Collie*, 79.

25. Bennett, *Collie*, 80.

26. Bennett, *Collie*, 97−106.

27. *Collie Folio*, 1 March 1906, 1.

28. H. Hunt, *Rough Collies: An Owner's Companion* (New York: Howell Book House, 1990), 68.

29. *Collie Folio*, November 1907, 142.

30. *Collie Folio*, 2 May 1906, 4.

31. *Collie Folio*, November 1907, 113.

32. *Collie Folio*, August 1909, 287.

33. *Collie Folio*, January 1910, 39.

34. *Collie Folio*, December 1909, 441.

35. *Collie Folio*, January 1911, 32.

36. *Collie Folio*, May 1912, 180.

37. *Collie Folio*, April 1913, 136.

38. G. Kaye, "W. E. Mason, Southport Collies," www.chelsea-collies.com/

southport.html 13, accessed September 1999; "A Century of Collies and Significant Developments: Part 1," *Collie Expressions*, January 2000, 33–35.

39. *Collie Folio*, 1 March 1906, 1.

40. *Canadian Poultry Review*, December 1877, 104.

41. Dalziel, *Collie: Its History*, 43.

42. *Farmer's Advocate*, 13 December 1906, 1932.

43. *Field and Fancy*, 11 August 1906, 12.

44. Palmer, "Scotch Collie."

45. *Collie Folio*, 1 March 1906, 1.

46. *Collie Folio*, March 1908, 83.

47. *Collie Folio*, March 1908, 84.

48. *Collie Folio*, November 1907, 140.

49. Jean Strouse's *Morgan: American Financier* (New York: Random House, 1999) shows the breadth of interests and the private nature of this extraordinary man but does not shed light on his breeding of Collies.

50. Morgan to Mead and Taft, 9 December 1892, Morgan Senior Letterpress Book 3 (1887–93), Pierpont Morgan Library Archives, New York.

51. *Collie Folio*, November 1907, 140.

52. *Collie Folio*, 1 December 1906, 39.

53. G. Kaye, "J. Pierpont Morgan," www.chelsea-collies.com/morgan.html, accessed 22 August 2000.

54. Morgan to Harrison, 11 June 1888, Morgan Senior Letterpress Book 3 (1887–93), Pierpont Morgan Library Archives, New York.

55. Morgan to Jarrett, 8 March 1893, Morgan Senior Letterpress Book 3 (1887–93), Pierpont Morgan Library Archives, New York.

56. Quoted in W. Stifel, *The Dog Show: 125 Years of Westminster* (New York: Westminster Kennel Club, 2001), 58–59.

57. Letter to the author from William Stifel, past president of the American Kennel Club, 25 April 2000. Stifel, *Dog Show*, 59.

58. *Collie Folio*, September 1910, 313.

59. *Collie Folio*, November 1907, 140.

60. *Collie Folio*, November 1907, 140.

61. Kaye, "J. Pierpont Morgan."

62. *Collie Folio*, June 1908, 181.

63. *Collie Folio*, November 1907, 140.

64. *Collie Folio*, June 1908, 181.

65. Robert Preston to J. P. Morgan Jr., 13 August 1918, J. P. Morgan Jr. Papers, box 105, Pierpont Morgan Library Archives, New York.

66. J. P. Morgan Jr.'s Secretary to R. Preston, 18 August 1918, J. P. Morgan Jr. Papers, box 105, Pierpont Morgan Library Archives, New York.

67. *Canadian Breeder and Agricultural Review*, 6 August 1885, 505.

68. *Kennel Gazette* (Canadian), August 1889, 53.

69. M. G. Denlinger, *The Complete Collie*, 3d ed. (Richmond, Va.: Denlinger's, 1949), 20.

70. Rawdon Lee, *A History and Description of the Collie or Sheep Dog in His British Varieties* (London: "The Field" Office, 1890), 17–18.

71. T. Garrison, "Out of the 'Gene' Pool?" *Collie Expressions*, July 1999, 36.

72. *Collie Folio*, January 1910, 7.

73. *Collie Folio*, February 1910, 48.

74. *Collie Folio*, February 1910, 49.

75. *Collie Folio*, March 1910, 84.

76. *Collie Folio*, April 1910, 140.

77. Coppinger and Coppinger, *Dogs*, 297–98.

78. O. Barnum, "Who Is Breeding the Old-Fashioned Collie?" *Country Life in America*, 15 December 1911, n.p.

79. R. A. Sturdevant, "The Case for the Modern Collie," *Country Life in America*, 1 March 1912, n.p.

80. "Save the Old-Fashioned Collie!" *Country Life in America*, 15 March 1912, n.p.

81. "The Old-Fashioned Collie," *Country Life in America*, 15 August 1912, n.p.

82. *Collie Folio*, May 1912, 167.

83. G. Gaye, "A Century of Collies and Significant Developments: Part 1," *Collie Expressions*, January 2000, 35.

84. "Friends of the Old Farm Collie, Bulletin 1," February 1995, izebug .syr.edu/~gsbisco/bull1a.htm, accessed 30 July 1999.

85. "Friends of the Old Farm Collie, Bulletin 2," October 1995, izebug .syr.edu/~gsbisco/bull2a.htm, accessed 30 July 1999.

86. "Friends of the Old Farm Collie, Bulletin 4," June 1997, izebug.syr .edu/~gsbisco/bull4a.htm, accessed 30 July 1999.

87. "Friends of the Old Farm Collie, Bulletin 5," December 1997, izebug .syr.edu/~gsbisco/bull5a.htm, accessed 30 July 1999.

88. "Classic Victorian Collie Club," izebug.syr.edu/~gsbisco/bull5b.htm, accessed 29 July 1999.

89. See, for example, "Old Fashioned Farm Collies at Faraway Farm," www.together.net/~silver61/, accessed 8 February 2001.

90. American Working Farmcollie Association, www.geocities.com/ farmcollie1, accessed 8 February 2001.

91. *Collie Folio*, November 1908.

92. *Collie Folio*, December 1908, 407.

93. *Collie Folio*, December 1908, 408.

94. *Collie Folio*, December 1908, 409.

95. *Collie Folio*, January 1909, 15.

96. *Collie Folio*, July 1909, 225.

97. *Collie Folio*, August 1909, 274.

98. J. H. Walsh ("Stonehenge"), *The Dogs of the British Islands* (London: "The Field" Office, 1882), 192–95.

99. *Canadian Poultry Review,* December 1877, 103.

100. Lee, *History and Description of the Collie,* 17.

101. E. C. Ash, *Dogs: Their History and Development,* vol. 1 (London: Ernest Benn, 1927), 275.

102. R. Leighton, *The New Dog Book: British Dogs and Their Foreign Relatives, with Chapters on Law, Breeding, Kennel Management, and Veterinary Treatment* (London: Cassell, 1907), 106; Bennett, *Collie,* 22; M. G. Denlinger, *The Complete Collie* (New York: Howell Book House, 1962), 9–11; Garrison, "Out of the 'Gene' Pool?" 36; I. Combe, D. Bridge, and P. Hutchinson, *Rough Collies of Distinction: A Pictorial Record of Influential British Rough Collies from the 1860s to the Present Day* (Cambridge: Combe, Bridge and Hutchinson, 2001), 39.

103. *Farmer's Advocate,* 14 April 1909, 540.

104. *Farmer's Advocate,* 16 January 1899, 36.

105. *Farming,* September 1895, 39.

106. K. L. Bates, *Sigurd, Our Golden Collie, and Other Companions of the Road* (London: J. M. Dent, 1921), 6.

107. Bates, *Sigurd,* 12.

108. Albert Payson Terhune, *Lad: A Dog* (New York: E. P. Dutton, 1919), 1.

109. K. Thomas, "Why You Should Remember Sunnybank Alton Andeen," *Collie Expressions,* April—May 1999, 31. For a detailed discussion of Terhune's breeding operations and even some of the pedigrees of his dogs, see K. Marshall, *His Dogs: Albert Payson Terhune and the Sunnybank Collies* (New York: Collie Club of America Foundation [Collie Health Foundation], 2001).

110. "The Dog Is a Business, Now," by Albert Payson Terhune, Albert Payson Terhune Papers, box 2, Library of Congress, Washington, D.C.

111. "The Collie: There Is Money in I Iim," by Albert Payson Terhune, Albert Payson Terhune Papers, box 2, Library of Congress, Washington, D.C.

112. "The Dog Is a Business, Now," by Albert Payson Terhune, Albert Payson Terhune Papers, box 2, Library of Congress, Washington, D.C.

113. I. Litvag, *The Master of Sunnybank: A Biography of Albert Payson Terhune* (New York: Harper and Row, 1977), 4.

114. Litvag, *Master of Sunnybank,* 82.

## *Five.* A World Market for Arabians Takes Shape

1. Lady Wentworth [Judith Blunt Lytton], *The Authentic Arabian Horse* (London: George Allen and Unwin, 1945), 27–28.

2. H. H. Reese, *The Kellogg Arabians: Their Background and Influence* (Alhambra, Calif.: Bordon, 1958), 101.

3. F. Knorr, "A History of the Arabian Horse and Its Influence on Mod-

ern Breeds," *American Breeders' Magazine 3* (1912): 174–75, 177; J. Forbis, *The Classic Arabian Horse* (New York: Liveright, 1976), 14.

4. Wentworth, *Authentic Arabian Horse*, 177.

5. J. Clutton-Brock, *Horse Power: Horse and Donkey in Human Societies* (London: Natural History Museum Publications, 1992), 61.

6. R. Archer, C. Pearson, and C. Covey, *The Crabbet Arabian Stud: Its History and Influence* (Northleach, U.K.: Alexander Heriot, 1978), 16; Wentworth, *Authentic Arabian Horse*, 160.

7. "Horse of the Desert Bedouin," Arabian Horse Registry of America, arabianhorseamerica.com/History/bedouin.asp, accessed 12 December 2001.

8. A few examples are as follows: Reese, *Kellogg Arabians*, 27, 30, 34; Forbis, *Classic Arabian Horse*, 254–55, 276; "Davenport Arabians at Craver Farms," in *Arabiana: An Anthology of Articles from "Your Pony" and "The International Rider and Driver,"* *1959 to 1974*, ed. W. Simpson (Fort Atkinson, Wis., 1975), 10–11; G. B. Edwards-Craver letters, in *Arabiana*, 15, 98–99; D. S. Whitman, "The Strain of It All," *American Arabian Online*, www.geocities.com/Heartland/Ranch/7485/strain.html 28, accessed August 1999; Wentworth, *Authentic Arabian Horse*, 143, 314; G. B. Edwards, "To Progress . . . or Regress: That Is the Question," *Arab Horse Journal* (hereafter *AHJ*), May 1960, 12–16; G. B. Edwards, "The Great Strain Robbery, or The Pursuit of the Pashas," Part 1, *AHJ*, August 1960, 22–24, 29; G. B. Edwards, "The Great Strain Robbery, or The Pursuit of the Pashas," Part 2, *AHJ*, September 1960, 22–26, 28; "Arabian Horses," *Canadian Breeder and Agricultural Review*, 8 April 1885, 214; E. B. Babcock and R. E. Clausen, *Genetics in Relation to Agriculture* (New York: McGraw-Hill, 1918), 444; E. Skorkowski, "Three Arabian Races," *AHJ*, June 1960, 30–32.

9. Forbis, *Classic Arabian Horse*, 254, 275.

10. M. Greely, *Arabian Exodus* (London: J. A. Allen, 1975), 32.

11. See Wilfrid Blunt, "Historic Sketch of the Rise and Decline of Wahhabism in Arabia," in *A Pilgrimage to Nejd*, by Lady Anne Blunt (1881; reprint, London: Century, 1985).

12. "Mohammed Ali and Abbas Pasha," *Arabian Horse World* (hereafter *AHW*), May 1982, 336–38, 343. Both the May 1981 and the May 1982 editions of the *AHW* have many articles on the history of the Arabian horse in relation to Egypt. See also Blunt, "Historic Sketch," and his preface to Blunt, *Pilgrimage to Nejd*.

13. "Arabian Types and Strains," in *Arabiana*, 252.

14. "Ali Pasha Sherif," *AHW*, May 1982, 343–44.

15. Greely, *Arabian Exodus*, 30–35; H. H. Reese, *Kellogg Arabians*, 39–40; Forbis, *Classic Arabian Horse*, 159; Edwards, "Great Strain Robbery, Part 2, 22–23; G. B. Edwards, "Glass Eyes and White Markings in Arabs," *AHJ*, April 1959, 20; M. Weise, "A Reference Guide: Arabian Color Coat," www.arabian-horses.com/feature/margo/color/, accessed 16 September 1999; J. Forbis, "The Arabian Horse in the Middle East," Part 3, *AHJ*, July 1960, 22.

16. Greely, *Arabian Exodus*, 44.

17. "Skowronek Type and the 'Raffles' Variation," in *Arabiana*, 103; "Comet," in *Arabiana*, 84.

18. M. J. Parkinson, "Skowronek," *AHW,* July 1982, 119.

19. Greely, *Arabian Exodus*, 170.

20. Archer, Pearson, and Covey, *Crabbet Arabian Stud*, 33. See also Lady Anne Blunt, *A Pilgrimage to Nejd* (1881; reprint, London: Century, 1985), and W. S. Blunt, *My Diaries*, 2 vols. (London: Martin Secher, 1900, 1914).

21. P. Upton, *Desert Heritage* (Cambridge: Burlington Press, 1980), 20, 36, 68, 80, 88.

22. "With the International," *Arabian Horse News*, May 1959, 12.

23. Archer, Pearson, and Covey, *Crabbet Arabian Stud*, 170.

24. G. B. Edwards, "Remember the Mare," *AHJ*, April 1961, 27.

25. Archer, Pearson, and Covey, *Crabbet Arabian Stud*, 47, 53, 54—55, 58, 60, 66, 75.

26. "Breeders of the Northeast: A Look Back," *AHW,* November 1988, 143—44.

27. "Breeders of the Northeast," 143—44; "Davenport Arabian Horses," in *Arabiana*, 1—5; M. Bowling, "The CMK Heritage," www.dsrtweyr.com/cmk/cmkmbheritage.html, accessed 28 August 1999; "Peter Bradley," *AHW,* July 1984, 447, 449—50.

28. R. Longrigg, *The History of Horse Racing* (London: Macmillan, 1972), 216, 219, 221—23, 233, 285.

29. "Breeders of the Northeast," 143—44; "Davenport Arabian Horses," in *Arabiana*, 1—5; Bowling, "CMK Heritage," accessed 28 August 1999.

30. "AHRA Importation Policies and History," WAHO (World Arabian Horse Organization) Publication 21, Part 4, www.euroarab.com/news/excerpt5 .htm, accessed 24 August 1999; "The Reason Behind the Registry," *American Arabian Online*, www.geocities.com/Heartland/Ranch/7485/registry.html, accessed 28 August 1999; "Mission and History," *Registry*, www.theregistry.org/mission/index.shtml, 22 December 2000.

31. "With the International," *Arabian Horse News*, May 1959, 12.

32. "Registration Certificates," *AHJ*, March 1960, 8.

33. *AHJ*, April 1960, 15; M. Weise, "*Witez II: The Look of Eagles," www .arabian-horses.com/feature/margo/color/black2:htm, accessed 21 September 1999.

34. Reese, *Kellogg Arabians*, 26.

35. Reese, *Kellogg Arabians*, 43.

36. Reese, *Kellogg Arabians*, 101.

37. Reese, *Kellogg Arabians*, 9, 13, 14, 17, 25.

38. "Ferzon: A Classic Gentleman," *AHW,* September 1987, 147; "The Skowronek Influence," *AHW,* August 1982, 97; "Skowronek," *AHW,* July 1982, 135; "Raffles: The Unlikely Legend," *AHW,* July 1982, 306, 311, 322; "Skowronek . . . Three in One," *AHJ*, February 1960, 26—28, 31—32.

39. Edwards, "Remember the Mare," 26.

40. Edwards, "Skowronek . . . Three to One," 26–28, 31–32.

41. "Non-Arab Blood," WAHO Publication 21, Part 3.

42. R. Archer, Pearson, and Covey, *Crabbet Arabian Stud*, 322.

43. *AHJ*, May 1960, 4, 16, 35, 34, 42; April 1960, 22, 41; August 1960, 2; December 1960, 31; May 1961, 33; April 1959, cover; October 1960, 12; March 1960, 21, February 1960, 19; December 1959; *AHW*, September 1987, 146.

44. P. Lindsay, "Arabian Breeding in Postwar Poland," *AHJ*, February 1961, 26.

45. *AHW*, June 1986, 427.

46. Lindsay, "Arabian Breeding in Postwar Poland," 29.

47. Weise, "*Witez II: The Look of Eagles," 6; "*Witez II and the Circle Two," *AHJ*, February 1961, 6–8.

48. *AHW*, June 1987, 349, 411.

49. *AHW*, April 1988, 145; July 1982, 368, 445; September 1987, 147.

50. *AHW*, August 1983, 112–13.

51. *AHW*, April 1982, 128.

52. *Arabian Horse News*, April 1958, 33.

53. *AHW*, April 1983, 184, 187, 189–90.

54. *AHW*, May 1980, 236, 237, 241; April 1983, 181, 183, 184, 187, 524; April 1984, 203.

55. *AHW*, August 1987, 191.

56. *AHW*, December 1987, 85.

57. *AHW*, April 1983, 190, 197, 201.

58. *AHW*, January 1986, 370.

59. *AHW*, June 1984, 518.

60. *AHW*, April 1983, 516, 279, 281, 282, 327, 328, 372, 396, 398, 403; December 1987, 53.

61. *AHW*, July 1985, 130–31; August 1986, 236–39.

62. *AHW*, June 1984, 514.

63. *AHW*, April 1983, 282, 327, 372.

64. *AHW*, November 1984, 46–47.

65. *AHW*, May 1982, 166, 447.

66. *AHW*, August 1981, AWH-3, AWH-5, AWH-13, AWH-17.

67. A. Magid, "History of the Egyptian Arabian in the United States," www.arabian-horse.com/feature/egyptian/index.html, accessed 28 August 1999; Forbis, *Classic Arabian*, 209, 253–54, 275–77.

68. *AHW*, August 1982, 13, 39.

69. *AHW*, May 1985, 309.

70. *AHW*, May 1981, 427, 429; July 1985, 241–42.

71. *AHW*, May 1981, 427.

72. *AHW*, December 1987, 53.

73. *AHW*, June 1984, 108.

74. *AHW,* January 1987, 931–95.

75. *AHW,* June 1984, 108–9.

76. *AHW,* November 1984, 81–121.

77. *AHW,* November 1985, 194, 195, 196, 201.

78. *AHW,* November 1984, 81–121; *Arabian* 1 (1975): 22.

79. *AHW,* November 1984, 323, 325, 336–37.

80. *AHW,* November 1985, 354–55.

81. *AHW,* April 1982, 177; February 1987, 255; September 1984, 263–64, 265–66.

82. *AHW,* December 1987, 16–17, 190; August 1983, 30–31; January 1987, 931–35.

83. J. Forbis, "Ansata Arabian Stud: The History," *AHW,* August 1982, 178; J. Forbis, "Ansata: The Breeding Program," *AHW,* August 1982, 202–3, 213.

84. "Raswan," in *Arabiana,* 148, "Arabian Types and Strains: 'Egyptian,' 'Blunt' and 'Crabbet,'" in *Arabiana,* 142, 145, 152; "Davenport Arabians at Craver Farms," in *Arabiana,* 10–11; G. B. Edwards, letter, in *Arabiana,* 15.

85. "Arabian Types and Strains," 142, 145, 152; "Blue Star Arabians, Blue List Arabians and Arabians," in *Arabiana,* 142, 145, 152.

86. "Davenport Arabians," in *Arabiana,* 3, 1198, 141.

87. See, for example, *AHW,* August 1983, 462–72.

88. *AHW,* August 1985, 293, 296, 297.

89. *AHW,* October 1985, 64.

90. *AHW,* May 1983, 386, 387, 389, 390, 392, 394, 402, 413.

91. *AHW,* May 1983, 394.

92. *AHW,* May 1985, 312.

93. *AHW,* September 1981, 243.

94. *AHW,* August 1982, 42; May 1982, 562; September 1981, 243.

95. *AHW,* June 1980, 185.

96. *AHW,* April 1986, 64, 374.

97. *AHW,* January 1986, 372.

98. *AHW,* March 1986, 56, 388, 389.

99. Note to the author from PricewaterhouseCoopers on the subject of American taxes, 18 January 2000. For a detailed review of how the tax act affected horse owners, see *AHW,* January 1988, 132–39.

100. *AHW,* December 1986, 84.

101. *AHW,* September 1987, 65.

102. *AHW,* April 1987, 116.

103. *AHW,* May 1988, 344–51.

104. *AHW,* September 1988, 110, 112.

105. *AHW,* April 1987, 119; January 1988, 454; April 1988, 121–25; December 1988, 15.

106. *AHW,* January 1988, 454, 455.

107. *AHW,* May 1987, 120–21.

108. *AHW,* September 1987, 387.

109. *AHW,* December 1988, 273.

110. *AHW,* September 1988, 160.

111. *AHW,* May 1988, 28–29.

112. "Ninety Years of Arabian Horses," *Registry News,* June 1998, www .theregistry.org/press/pr970526.shtml, accessed 24 August 1999.

113. K. Berkery, "Twenty Years after Bask," www.arabian-horses.com/feature /bask/index.html, accessed 31 August 1999.

## *Six.* The Arabian Horse Registry of America

1. "Mission and History," *The Registry,* www.theregistry.org/mission/index .shtml, accessed 25 August 1999.

2. *Arabian Horse World* (hereafter *AHW* ), June 1987, 410.

3. *Arab Horse Journal* (hereafter *AHJ* ), March 1961, 31, 37.

4. "AHRA Importation Policies and History," WAHO Publication 21, Part 4, www.euroarab.com/news/excerpt4.htm, accessed 24 August 1999. This document is supported by evidence from many archival sources.

5. "Purity Wasn't the Issue at All," WAHO Publication 21, Part 5, www .waho.org/Purityissue.html, accessed 28 August 1999.

6. *AHW,* May 1982, 160, 433, 447–48.

7. "AHRA Importation Policies and History," accessed 28 August 1999, 12 December 2001. Ideas here are supported by many other sources.

8. *AHW,* August 1985, 290; "Purity Wasn't the Issue at All," accessed 25 August 1999.

9. *AHW,* September 1980, 241, 243; August 1985, 289–90.

10. *AHW,* July 1986, 404–5.

11. *Arabian,* 1975, 7.

12. *AHW,* April 1986, 237, 240, 241.

13. *AHW,* March 1987, 191, 194.

14. *AHW,* January 1987, 931–35.

15. *AHW,* March 1987, 191, 194.

16. *Arabian,* 1974, 17.

17. *Arabian,* 1974, 17.

18. *Arabian,* 1974, 18.

19. *Arabian,* 1974, 18.

20. *AHW,* August 1985, 290.

21. *AHW,* July 1986, 406.

22. "Is Purity the Issue?" accessed 26 September 1999. Many sections of this long document are useful.

23. "General Information—Is Purity the Issue?" WAHO Publication 21, Part 4, www.euroarab.com/news/excerpt4.htm, accessed 26 September 1999.

24. G. B. Edwards, "The Great Strain Robbery, or The Pursuit of the Pashas," Part 2, *AHJ,* September 1960, 22.

25. Lady Wentworth, *The Authentic Arabian Horse* (London: George Allen and Unwin, 1945), 307–9.

26. Quoted in full in M. Greely, *Arabian Exodus* (London: J. A. Allen, 1975), 104.

27. "Non-Arab Blood in AHRA Pedigrees—the Evidence," WAHO Publication 21, Part 3, euroarab.com/news/excerpt3.htm, accessed 24 August 1999.

28. "Non-Arab Blood," accessed 24 August 1999.

29. "Non-Arab Blood," accessed 26 September 1999.

30. "Summary—Definition or Deception," WAHO Publication 21, Part 1, www.waho.org/Purityissue.html, accessed 11 December 2001.

31. "AHRA Importation Polices and History," WAHO Publication 21, Part 4, www.euroarab.com/news/excerpt4.htm, accessed 26 September 1999.

32. "WAHO Threatens US Arabian Breed Purity," press release of *Registry*, December 1996, www.theregistry.org/press/pr9612.shtml, accessed 23 August 1999.

33. "AHRA Refuses WAHO Definition of Breed Purity," press release of *Registry*, January 1997, www.theregistry.org/press/pr970124.shtml, accessed 24 August 1999.

34. "Registry Expelled from WAHO," *Registry News*, December 1997, www.theregistry.org/registry/dec97/expelled.shtml, accessed 24 August 1999.

35. *AHW*, August 1984, 472; August 1985, 290.

36. "The Resignation of Canada (CAHR) from WAHO Membership," www.waho.org/News.html, accessed 26 September 1999.

37. "AHRA and CAHR Reaffirm Support," press release of *Registry*, 16 March 1998, www.theregistry.org/press/pr980316.shtml, accessed 24 August 1999.

38. Letter, euroarab.com/news/aha/htm, accessed 24 August 1999.

39. "AHRA and Jordan Sign . . . Agreement," press release of *Registry*, 13 February 1998, www.theregistry.org/press/pr980213.shtml, accessed 24 August 1999.

40. "WAHO vs. AHRA," in "The Many Faces of WAHO," *Arabian Horse Interactive Forum*, www.arabian-horses.com/feature/iforum/forum3.htm, accessed 24 August 1999.

41. Press release of *Registry*, 29 January 1999, www.theregistry.org/press/pr990129.shtml, accessed 14 December 2000.

42. "Questions and Answers—Alliance of the Americas," press release, 6 October 2001, *Registry*, www.theregistry.org/News/PressReleasesView.asp?id=11, accessed 12 December 2001.

43. *AHW*, January 2003, 400. See also the website of the IAHA and the AHRA.

## Concluding Remarks

1. M. G. Denlinger, *The Complete Collie*, 3d ed. (Richmond, Va.: Denlinger's, 1949), 55.

# ESSAY ON SOURCES

The basic structure of registries, tariffs relating to animals, and trade regulations pertaining to purebred stock—all of which proved critical to the way purebred breeding worked—emerged from government documents. Agricultural departments reported on any government action that affected the production of animals. While government documents also provided information on animal breeding generally, they were invaluable for a study of Shorthorn breeding. Specialized material yielded more information on Shorthorns. For example, for a set of published letters between a Scottish breeder of Shorthorns, Amos Cruickshank, and an Ontario farmer, John I. Davidson, see T. B. Marson, *The Shorthorns of Scotland: Sittyton* (Edinburgh: Scottish Shorthorn Breeders' Association, 1948). The letters date from 1873 to 1891. Examples of other primary book material used to study Shorthorns in particular were I. M. Bruce, *The History of the Aberdeenshire Shorthorn* (Aberdeen: Aberdeen, Banff and Kincardine Shorthorn Breeders' Association, 1923); *The Canada Herd Book, Containing the Pedigrees of Improved Short-Horned Cattle*, vol. 1 (Toronto: Board of Agriculture of Canada West, 1867); *History of Short-horned Cattle Imported into the Present Dominion of Canada from Britain and United States, Chronologically Arranged*, vols. 1–10 (Toronto: Agriculture and Arts Association, 1895); W. Housman, *The Improved Shorthorn: Notes and Reflections upon Some Facts in Shorthorn History, with Remarks upon Certain Principles of Breeding* (London: Ridgeway, 1876); T. B. Marson, *Scotland, the World's Stud Shorthorn Farm: A Short Account of Some of Scotland's Famous Herds* (Edinburgh: Scottish Shorthorn Breeders' Association, 1937); A. Sanders, *Short-Horn Cattle: A Series of Historical Sketches, Memoirs and Records of the Breed and Its Development in the United States and Canada* (Chicago: Sanders, 1900); A. Sanders, *Red, White and Roan* (Chicago: American Shorthorn Breeders' Association, 1936); J. Sinclair, *History of Shorthorn Cattle* (London: Vinton, 1907).

For Collies some good primary book sources were H. Dalziel, *British Dogs: Describing the History, Characteristics, Points, and Club Standards of the Various Breeds of Dogs Established in Great Britain*, 2d ed., vol. 2 (London: L. Upcott Gill, 1889); O. P. Bennett and T. M. Halpin, *The Collie* (Washington, Ill.: O. P. Bennett, 1942); O. P. Bennett and C. H. Wheeler, *The Collie* (Washington, Ill.: O. P. Bennett, 1924); I. Combe, D. Bridge, and P. Hutchinson, *Rough Collies of Distinction: A Pic-*

*torial Record of Influential British Rough Collies from the 1860s to the Present Day* (Cambridge: I. Combe, D. Bridge, and P. Hutchinson, 2001); H. Dalziel, *The Collie: As a Show Dog, Companion, and Worker,* 4th ed., rev. J. Maxtee (London: Bazaar, Exchange and Mart Office, 1921); H. Dalziel, *The Collie: Its History, Points, and Breeding* (London: L. Upcott Gill, 1888); M. G. Denlinger, *The Complete Collie* (Richmond, Va.: Denlinger's, 1949); M. G. Denlinger, *The Complete Collie* (New York: Howell Book House, 1962); Collie Club of America, *The New Collie* (New York: Howell Book House, 1983); J. H. Walsh ("Stonehenge"), *The Dogs of the British Islands* (London: Field Office, 1882); J. Watson, *The Dog Book: A Popular History of the Dog, with Practical Information as to Care and Management of House, Kennel, and Exhibition Dogs; and Descriptions of All the Important Breeds* (Toronto: Musson, 1906).

For Arabians primary book sources to pursue are R. Archer, C. Pearson, and C. Covey, *The Crabbet Arabian Stud: Its History and Influence* (Northleach, U.K.: Alexander Heriot, 1978); J. Forbis, *The Classic Arabian Horse* (New York: Liveright, 1976); C. Raswan, *The Arab and His Horse* (Oakland, Calif.: Raswan, 1955); H. H. Reese, *The Kellogg Arabians: Their Background and Influence* (Alhambra, Calif.: Bordon, 1958); Lady Wentworth [Judith Blunt Lytton], *The Authentic Arabian Horse* (London: George Allen and Unwin, 1945); and Lady Wentworth, *Thoroughbred Racing Stock* (New York: Charles Scribner's Sons, 1938).

Good supportive primary book sources for subjects related to Shorthorns, Collies, and Arabians were Lady Anne Blunt, *A Pilgrimage to Nejd* (1881; reprint, London: Century, 1985); W. S. Blunt, *My Diaries,* vol. 1 [1888–1900] and vol. 2 [1900–1914] (London: Martin Secher, 1900, 1914); A. E. Ash, *Dogs: Their History and Development,* 2 vols. (London: Ernest Benn, 1927); G. P. Burnham, *The History of the Hen Fever* (1855; reprint San Diego: Frank E. Marcy, 1935); R. W. Dickson, *Improved Live Stock and Cattle Management* (London: Thomas Kelly, 1825); H. B. Hawes, *Dogs and a Suggestion: Address by Harry B. Hawes under the Auspices of the State Board of Agriculture* (Columbia: University of Missouri, 1914); H. B. Hawes, *The Dog: Remarks of Hon. Harry B. Hawes in the House of Representatives* (Washington, D.C.: Government Printing Office, 1923); E. W. Jaquet, *The Kennel Club: A History and Record of Its Work* (London: Kennel Gazette, 1905); R. Lee, *A History and Description of the Collie or Sheep Dog in His British Varieties* (London: "The Field" Office, 1890); R. Leighton, *The New Dog Book: British Dogs and Their Foreign Relatives, with Chapters on Law, Breeding, Kennel Management, and Veterinary Treatment* (London: Cassell, 1907); J. Lush, "Notes on Animal Breeding," manuscript, 1933, University of Guelph; J. Blunt Lytton [Lady Wentworth], *Toy Dogs and Their Ancestors* (London: Duckworth, 1911); D. Marshall, *Shorthorn Cattle in Canada* ([Toronto]: Dominion Shorthorn Breeders' Association, 1932); "Minutes of Evidence Taken before the Departmental Committee Appointed to Enquire and Report as to the British Trade in Live Stock with the Colonies and Other Countries," 1912, Cd. 6032 (Britain); H. E. Packwood, *Show Collies: Rough and Smooth Coated, a Complete History* (n.p., 1906); "Proceedings of the Fourth Annual Convention of the American Association of Breeders of Short Horns,"

Toronto, 1875 (Toronto: Globe, 1875); J. H. Sanders, *The Breeds of Live Stock, and the Principles of Heredity* (Chicago: J. H. Sanders, 1887); W. Simpson, ed., *Arabiana: An Anthology of Articles from "Your Pony" and "The International Rider and Driver," 1959 to 1974* (Fort Atkinson, Wis., 1975); J. H. Walsh ("Stonehenge"), *British Rural Sports* (London: Frederick Warne, 1875).

Journals provided extremely valuable primary material, in fact probably the most important for the book. The most significant were the *Arabian*, the *Arab Horse Journal*, the *Arabian Horse News*, the *Arab Horse Society News*, *Arabian Horse World*, the *Canadian Poultry Review*, *Collie Cues*, the *Collie Folio*, *Collie Reflections*, *Dog Fancy*, *Farmer's Advocate*, *Field and Fancy*, *Kennel Gazette* (Canadian), the *Kennel Gazette* (British), the *Illustrated London News*, *Pure-Bred Dogs: American Kennel Gazette*, the *Ladies' Kennel Journal*, and *American Breeders' Magazine* (later the *Journal of Heredity*).

Valuable manuscript collections I looked at were Arthur Johnston Papers, Ontario Archives; Miller-Davidson Papers (where original Cruickshank letters that have not been published can be found), University of Guelph Archives; J. P. Morgan Letterpresses and Louisa Morgan Satterlee Photo Albums, Pierpont Morgan Library; and Albert Payson Terhune Papers, Library of Congress. My other major primary sources arose from the Internet. In many ways it served the role of a modern journal, reflecting what people think and also how they communicate about issues that concerned them. The immediacy of the Farmcollie movement or of the WAHO-AHRA debate gave me a sense of excitement and participation. The Internet proves a new and challenging primary source for historians.

Secondary sources yielded material on specialized subjects (such as the history of kennel clubs) and also gave background information on society, culture, and economics as well as on the domestication of animals, early selective breeding, agriculture, and genetics. These secondary sources reflect the work of humanist and scientific scholars. Some books particularly worth looking at are D. Barnes, ed., *The AKC's World of the Pure-Bred Dog* (New York: Howell Book House, 1983); J. Beckett, *The Agricultural Revolution* (Cambridge, Mass.: Basil Blackwell, 1990); R. A. Caras, *Going for the Blue: Inside the World of Show Dogs and Dog Shows* (New York: Warner Books, 2001); J. Clutton-Brock, *Domesticated Animals from Early Times* (Heinemann: British Museum [Natural History], 1981); J. Clutton-Brock, *A Natural History of Domesticated Mammals*, 2d ed. (Cambridge: Cambridge University Press, 1999); R. Coppinger and L. Coppinger, *Dogs* (New York: Scribner, 2001); I. Crawford-Siano, *Journey to Perfection: The Agricultural Art of Ross Butler* (Kingston, Ont.: Quarry Press, 1997); M. Derry, *Ontario's Cattle Kingdom: Purebred Breeders and Their World* (Toronto: University of Toronto Press, 2001); A. Fraser, *Animal Husbandry Heresies* (London: Crosby Lockwood, 1960); T. Grandin, ed., *Genetics and the Behaviour of Domestic Animals* (London: Academic Press, 1998); P. Henlein, "Cattle Kingdom in the Ohio Valley: The Beef Cattle Industry in the Ohio Valley, 1783–1860" (Ph.D. diss., University of Wisconsin, 1957); W. C. Hill and T. F. C. Mackay, eds., *Evolution and Animal Breeding:*

*Reviews on Molecular and Quantitative Approaches in Honour of Alan Robertson* (Wallingford, U.K.: C. A. B. International, 1989); M. Huggins, *Flat Racing and British Society, 1790–1914* (London: Frank Cass, 2000); F. Jackson, *Crufts: The Official History* (London: Pelham Books, 1990); P. C. Johnson, *Farm Animals in the Making of America* (Des Moines, Iowa: Wallace Homestead Book Company, 1975); Kennel Club, *Treasures of the Kennel Club: Paintings, Personalities, Pedigrees and Pets* (London: Kennel Club, 2000); M. Lerner and H. Donald, *Modern Developments in Animal Breeding* (New York: Academic Press, 1966); H. F. Lionberger, *Adoption of New Ideas and Practices* (Ames: Iowa State Press, 1960); I. Litvag, *The Master of Sunnybank: A Biography of Albert Payson Terhune* (New York: Harper and Row, 1977); K. Marshall, *His Dogs: Albert Payson Terhune and the Sunnybank Collies* (New York: Collie Club of America Foundation [Collie Health Foundation], 2001); A. E. Mourant and F. E. Zeuer, eds., *Man and Cattle* (n.p.: Royal Anthropological Institute of Great Britain and Ireland, 1963); E. Longford, *A Pilgrimage of Passion: The Life of Wilfrid Scawen Blunt* (London: Weidenfeld and Nicolson, 1979); R. Longrigg, *The History of Horse Racing* (London: Macmillan, 1972); Harriet Ritvo's two books, *The Animal Estate* (Cambridge: Harvard University Press, 1987) and *The Platypus and the Mermaid and Other Figments of the Classifying Imagination* (Cambridge: Harvard University Press, 1997); N. Russell, *Like Engend'ring Like: Heredity and Animal Breeding in Early Modern England* (Cambridge: Cambridge University Press, 1986); G. MacEwan, *Heavy Horses: Highlights of Their History,* (Saskatoon, Sask.: Western Producer Prairie Books, 1986); G. MacEwan, *Highlights of Shorthorn History* (Winnipeg: Hignill, 1982); P. M. H. Mazumdar, *Eugenics, Human Genetics and Human Failings: The Eugenics Society, Its Sources and Its Critics in Britain* (London: Routledge, 1992); E. W. Minchinton, ed., *Essays in Agrarian History,* vols. 1 and 2 (Newton Abbot, U.K.: David and Charles, 1968); G. Mingay and J. D. Chambers, *The Agricultural Revolution, 1750–1880* (London: Batsford, 1966); E. Montcrieff, S. Joseph, and I. Joseph, *Farm Animal Portraits* (Woodbridge, U.K.: Antique Collectors' Club, 1996); W. Secord, *A Breed Apart: The Art Collection of the American Kennel Club and the American Kennel Club Museum of the Dog* (Woodbridge, U.K.: Antique Collectors' Club, 2001); W. Secord, *Dog Painting, 1820–1940: A Social History of the Dog in Art* (Woodbridge, U.K.: Antique Collectors' Club, 1992); W. Secord, *Dog Painting: The European Breeds* (Woodbridge, U.K.: Antique Collectors' Club, 2000); J. Serpell, *In the Company of Animals: A Study of Human-Animal Relations* (Cambridge: Cambridge University Press, 1996); J. Serpell, ed., *The Domestic Dog: Its Evolution, Behaviour, and Interactions with People* (Cambridge: Cambridge University Press, 1995); W. Stifel, *The Dog Show: 125 Years of Westminster* (New York: Westminster Kennel Club, 2001); C. G. Sutton, *Dog Breeding and Showing* (New York: Arco, 1983); K. Thomas, *Man and the Natural World: Changing Attitudes in England, 1500–1800* (London: Allen Lane, 1983); R. Trow-Smith, *A History of British Livestock Husbandry, 1700–1900* (London: Routledge and Kegan Paul, 1959); and K. Unkelbach, *Albert Payson Terhune, the Master of Sunnybank* (New York: Charterhouse, 1972).

Many wonderful articles, too numerous to list, can be found particularly in the two major agricultural journals, *Agricultural History* and *Agricultural History Review*. A few—from these journals and others—might be pointed out. John Walton's articles should be mentioned: "The Diffusion of Improved Short-horn Cattle in Britain during the Eighteenth and Nineteenth Centuries," *Transactions of the Institute of British Geographers*, n.s., 9 (1984): 22–36; "Pedigree and Productivity in the British and North American Cattle Kingdoms before 1930," *Journal of Historical Geography* 25 (1999): 441–62; "Pedigree and the National Cattle Herd circa 1750–1950," *Agricultural History Review* 34 (1986): 149–70. Some other articles worth mentioning here are R. F. Moore-Colyer, "Gentlemen, Horses and the Turf in Nineteenth Century Wales," *Welsh Historical Review* 16 (1992); R. F. Moore-Colyer, "Aspects of Horse Breeding and the Supply of Horses in Victorian Britain," *Agricultural History Review* 43 (1995): 47–60; R. F. Moore-Colyer, "Aspects of the Trade in British Pedigree Draught Horses with the United States and Canada, c. 1850–1920," *Agricultural History Review* 48 (2000): 42–59; M. Huggins, "Thoroughbred Breeding in the North and East Ridings of Yorkshire in the Nineteenth Century," *Agricultural History* 52 (1978): 247–61; H. Ritvo, "Pride and Pedigree: The Evolution of the Victorian Dog Fancy," *Victorian Studies* 29 (1986): 227–53; E. A. Wasson, "The Third Earl of Spencer and Agriculture, 1818–1845," *Agricultural History Review* 26 (1978): 89–99; and E. Whetham, "The Trade in Pedigree Livestock, 1850–1910," *Agricultural History Review* 27 (1978): 47–50. Also see K. White, "Victorian and Edwardian Dogs," *Veterinary History*, n.s., 7 (1992): 72–78.

A few notes on how I used these sources are in order. Fundamental to this book was an understanding of the animals themselves; that entailed an appreciation of certain breeds, and within those breeds certain bloodlines made evident in pedigrees. I studied pedigrees and came to know the influence of many individual animals on those pedigrees. I did so by examining books written by breeders on the significant individuals of the breed, breed journals, reports of breed associations, and pedigrees themselves. This process made it clear that single animals often could explain entire patterns. By reading books by different breeders of the same breed who wrote at different times about the same lines of breeding and the same animals, I became aware of various perspectives on certain problems. I could see the ideology behind the breeding technique and what influenced the popularity of certain lines of stock in light of the ideal. Breed journals provided useful information on techniques, attitudes toward techniques, and the results produced by the techniques. These journals contained letters from various people who expressed differing opinions about issues that concerned breeders at particular points in time. I thus gained a fairly broad picture of events surrounding breeding programs through a combined look at these specialized books and breed journals.

Letters and manuscript collections supplied another dimension to this story. So did some of the earliest manuals on how to breed animals. Government documents relating to agriculture provided a surprising amount of material on breeding, not just of farm animals but also of dogs and Arabian horses. Because of the critical role that agriculture played, general monographs on the economy and social structure of farming from the late eighteenth century to the twentieth century gave contextual background to all the material. So did books by biologists, geneticists, and students of animal behavior.

# INDEX